Lecture Notes in Control and Information Sciences

Edited by M. Thoma and A. Wyner

88

Bruce A. Francis

A Course in H_∞ Control Theory

Springer-Verlag
Berlin Heidelberg New York
London Paris Tokyo

Author
Prof. Bruce A. Francis
Dept. of Electrical Engineering
University of Toronto
Toronto, Ontario
Canada M5S 1A4

ISBN 3-540-17069-3 Springer-Verlag Berlin Heidelberg New York
ISBN 0-387-17069-3 Springer-Verlag New York Berlin Heidelberg

Offsetprinting: Mercedes-Druck, Berlin
Binding: B. Helm, Berlin
2161/3020-543210

To my parents

PREFACE

My aim in this book is to give an elementary treatment of linear control theory with an $\mathbf{H}_\infty$ optimality criterion. The systems are all linear, time-invariant, and finite-dimensional and they operate in continuous time. The book has been used in a one-semester graduate course, with only a few prerequisites: classical control theory, linear systems (state-space and input-output viewpoints), and a bit of real and complex analysis.

Only one problem is solved in this book: how to design a controller which minimizes the $\mathbf{H}_\infty$-norm of a pre-designated closed-loop transfer matrix. The $\mathbf{H}_\infty$-norm of a transfer matrix is the maximum over all frequencies of its largest singular value. In this problem the plant is fixed and known, although a certain robust stabilization problem can be recast in this form. The general robust performance problem – how to design a controller which is $\mathbf{H}_\infty$-optimal for the worst plant in a pre-specified set – is as yet unsolved.

The book focuses on the mathematics of $\mathbf{H}_\infty$ control. Generally speaking, the theory is developed in the input-output (operator) framework, while computational procedures are presented in the state-space framework. However, I have compromised in some proofs: if a result is required for computations and if both operator and state-space proofs are available, I have usually adopted the latter. The book contains several numerical examples, which were performed using PC-MATLAB and the Control Systems Toolbox. The primary purpose of the examples is to illustrate the theory, although two are examples of (not entirely realistic) $\mathbf{H}_\infty$ designs. A good project for the future would be a collection of case studies of $\mathbf{H}_\infty$ designs.

Chapter 1 motivates the approach by looking at two example control problems: robust stabilization and wideband disturbance attenuation. Chapter 2 collects some elementary concepts and facts concerning spaces of functions, both time-domain and frequency domain. Then the main problem, called the standard problem, is posed in Chapter 3. One example of the standard problem is the model-matching problem of designing a cascade controller to minimize the error between the input-output response of a plant and that of a model. In Chapter 4 the very useful parametrization due to Youla, Jabr, and Bongiorno (1976) is used to reduce the standard problem to the model-matching problem. The results in Chapter 4 are fairly routine generalizations of those in the expert book by Vidyasagar (1985a).

Chapter 5 introduces some basic concepts about operators on Hilbert space and presents some useful facts about Hankel operators, including Nehari's theorem. This material permits a solution to the scalar-valued model-matching problem in Chapter 6. The matrix-valued problem is much harder and requires a preliminary chapter, Chapter 7, on factorization theory. The basic factorization theorem is due to Bart, Gohberg, and Kaashoek (1979); its application yields spectral factorization, inner-outer factorization, and J-spectral factorization. This arsenal together with the geometric theory of Ball and Helton (1983) is used against the matrix-valued problem in Chapter 8; actually, only nearly optimal solutions are derived.

Thus Chapters 4 to 8 constitute a theory of how to compute solutions to the standard problem. But the $\mathbf{H}_\infty$ approach offers more than this: it yields qualitative and quantitative results on achievable performance, showing the trade-offs involved in frequency-domain design. Three examples of such results are presented in the final chapter.

I chose to omit three elements of the theory: a proof of Nehari's theorem, because it would take us too far afield; a proof of the main existence theorem, for the same reason; and the theory of truly (rather than nearly) optimal solutions, because it's too hard for an elementary course.

It is a pleasure to express my gratitude to three colleagues: George Zames, Bill Helton, and John Doyle. Because of George's creativity and enthusiasm I became interested in the subject in the first place. From Bill I learned some beautiful operator theory. And from John I learned "the big picture" and how to compute using state-space methods. I am also grateful to John for his invitation to participate in the ONR/Honeywell workshop (1984). The notes from that workshop led to a joint expository paper, which led in turn to this book.

I am also very grateful to Linda Espeut for typing the first draft into the computer and to John Hepburn for helping me with unix, troff, pic, and grap.

Toronto
May, 1986

Bruce A. Francis

SYMBOLS

$\mathbf{R}$	field of real numbers
$\mathbf{C}$	field of complex numbers
$\mathbf{L}_2(-\infty, \infty)$	time-domain Lebesgue space
$\mathbf{L}_2(-\infty, 0]$	ditto
$\mathbf{L}_2[0, \infty)$	ditto
$\mathbf{L}_2$	frequency-domain Lebesgue space
$\mathbf{L}_\infty$	ditto
$\mathbf{H}_2$	Hardy space
$\mathbf{H}_\infty$	ditto
prefix $\mathbf{R}$	real-rational
$\lVert\cdot\rVert$	norm on $\mathbf{C}^{n \times m}$; maximum singular value
$\lVert\cdot\rVert_2$	norm on $\mathbf{L}_2$
$\lVert\cdot\rVert_\infty$	norm on $\mathbf{L}_\infty$
superscript $\perp$	orthogonal complement
A^T	transpose of matrix A
A^*	complex-conjugate transpose of matrix A
Φ^*	adjoint of operator Φ
$F^\sim(s)$	$F(-s)^T$
Π_1	orthogonal projection from $\mathbf{L}_2$ to $\mathbf{H}_2^\perp$
Π_2	orthogonal projection from $\mathbf{L}_2$ to $\mathbf{H}_2$
Γ_F	Hankel operator with symbol F
$\Phi\mathbf{X}$	image of $\mathbf{X}$ under Φ
Im	image
Ker	kernel
$\mathbf{X}_-(A)$	stable modal subspace relative to A
$\mathbf{X}_+(A)$	unstable modal subspace relative to A

The transfer matrix corresponding to the state-space realization (A, B, C, D) is denoted $[A, B, C, D]$, i.e.

$$[A, B, C, D] := D + C(s - A)^{-1}B .$$

Following is a collection of useful operations on transfer matrices using this data structure:

$$[A, B, C, D] = [T^{-1}AT, T^{-1}B, CT, D]$$

$$[A, B, C, D]^{-1} = [A - BD^{-1}C, BD^{-1}, -D^{-1}C, D^{-1}]$$

$$[A, B, C, D]^{\sim} = [-A^T, -C^T, B^T, D^T]$$

$$[A_1, B_1, C_1, D_1] \times [A_2, B_2, C_2, D_2]$$
$$= \left[\begin{bmatrix} A_1 & B_1C_2 \\ 0 & A_2 \end{bmatrix}, \begin{bmatrix} B_1D_2 \\ B_2 \end{bmatrix}, [C_1 \quad D_1C_2], D_1D_2 \right]$$
$$= \left[\begin{bmatrix} A_2 & 0 \\ B_1C_2 & A_1 \end{bmatrix}, \begin{bmatrix} B_2 \\ B_1D_2 \end{bmatrix}, [D_1C_2 \quad C_1], D_1D_2 \right]$$

$$[A_1, B_1, C_1, D_1] + [A_2, B_2, C_2, D_2]$$
$$= \left[\begin{bmatrix} A_1 & 0 \\ 0 & A_2 \end{bmatrix}, \begin{bmatrix} B_1 \\ B_2 \end{bmatrix}, [C_1 \quad C_2], D_1 + D_2 \right]$$

CONTENTS

CHAPTER 1

INTRODUCTION

This course is about the design of control systems to meet frequency-domain performance specifications. This introduction presents two example problems by way of motivating the approach to be developed in the course. We shall restrict attention to single-input, single-output systems for simplicity.

To begin, we need the Hardy space $\mathbf{H}_\infty$. This consists of all complex-valued functions $F(s)$ of a complex variable s which are analytic and bounded in the open right half-plane, Re $s>0$; bounded means that there is a real number b such that

$$|F(s)| \leq b, \quad \text{Re } s > 0 .$$

The least such bound b is the $\mathbf{H}_\infty$-*norm* of F, denoted $\|F\|_\infty$. Equivalently

$$\|F\|_\infty := \sup \{ |F(s)| : \text{Re } s > 0 \} . \tag{1}$$

Let's focus on real-rational functions, i.e. rational functions with real coefficients. The subset of $\mathbf{H}_\infty$ consisting of real-rational functions will be denoted by $\mathbf{RH}_\infty$. If $F(s)$ is real-rational, then $F \in \mathbf{RH}_\infty$ if and only if F is *proper* ($|F(\infty)|$ is finite) and *stable* (F has no poles in the closed right half-plane, Re $s \geq 0$). By the maximum modulus theorem we can replace the open right half-plane in (1) by the imaginary axis:

$$\|F\|_\infty = \sup \{ |F(j\omega)| : \omega \in \mathbf{R} \} . \tag{2}$$

To appreciate the concept of $\mathbf{H}_\infty$-norm in familiar terms, picture the Nyquist plot of $F(s)$. Then (2) says that $\|F\|_\infty$ equals the distance from the origin to the farthest point on the Nyquist plot.

We now look at two examples of control objectives which are characterizable as $\mathbf{H}_\infty$-norm constraints.

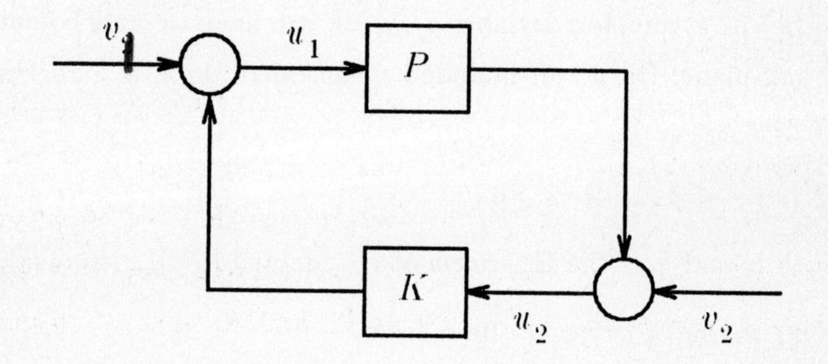

Figure 1.1. Single-loop feedback system

Example 1.

The first example uses a baby version of the small gain theorem. Consider the feedback system in Figure 1. Here $P(s)$ and $K(s)$ are transfer functions and are assumed to be real-rational, proper, and stable. For well-posedness we shall assume that P or K (or both) is *strictly proper* (equal to zero at $s=\infty$). The feedback system is said to be *internally stable* if the four transfer functions from v_1 and v_2 to u_1 and u_2 are all stable (they are all proper because of the assumptions on P and K). For example, the transfer function from v_1 to u_1 equals $(1-PK)^{-1}$. The Nyquist criterion says that the feedback system is internally stable if and only if the Nyquist plot of PK doesn't pass through or encircle the point $s=1$. So a sufficient condition for internal stability is the small gain condition $||PK||_\infty<1$.

Let's extend this idea to the problem of robust stabilization. The block diagram in Figure 2a shows a plant and controller with transfer functions $P(s)+\Delta P(s)$ and $K(s)$ respectively; P represents the nominal plant and ΔP an unknown perturbation, usually due to unmodeled dynamics or parameter variations. Suppose, for simplicity, that P, ΔP, and K are real-rational, P and ΔP are strictly proper and stable, and K is proper. Suppose also that the feedback system is internally stable for $\Delta P=0$. How large can $|\Delta P|$ be so that internal stability is maintained?

One method which is used to obtain a transfer function model is a frequency response experiment. This yields gain and phase estimates at several frequencies, which in turn provide an upper bound for $|\Delta P(j\omega)|$ at several values of ω. Suppose R is a radius function belonging to $\mathbf{RH}_\infty$ and bounding the perturbation ΔP in the sense that

$$|\Delta P(j\omega)| < |R(j\omega)| \text{ for all } 0\leq\omega\leq\infty ,$$

or equivalently

$$||R^{-1}\Delta P||_\infty < 1 . \tag{3}$$

How large can $|R|$ be so that internal stability is maintained?

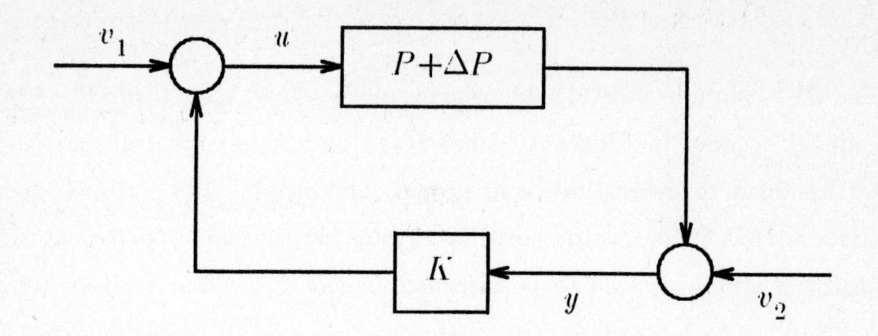

Figure 1.2a. Feedback system with perturbed plant

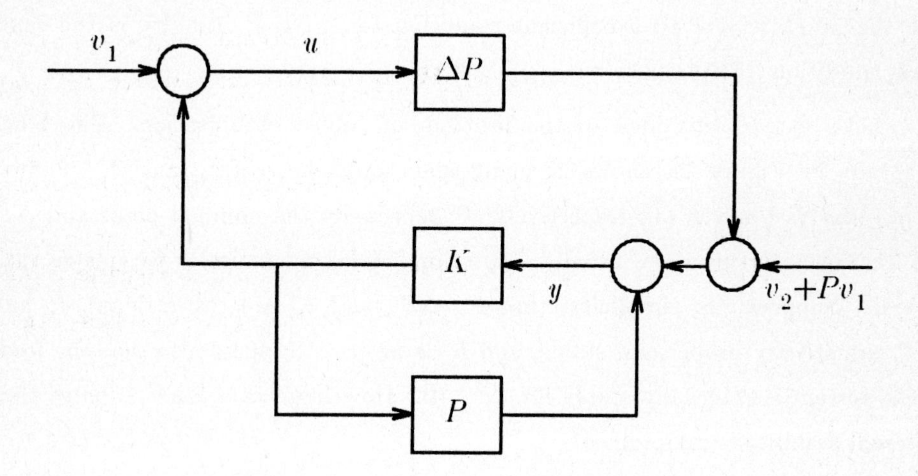

Figure 1.2b. After loop transformation

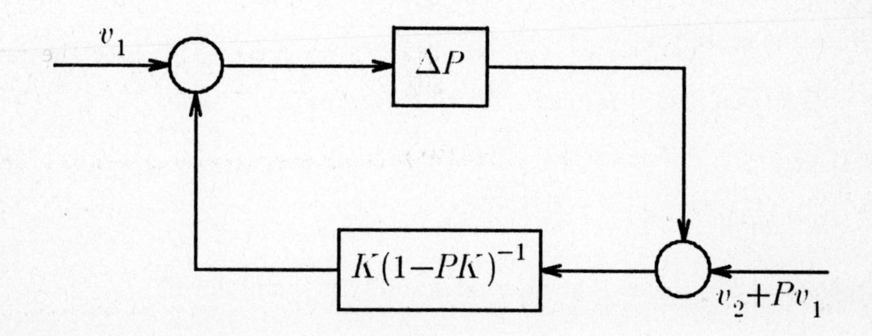

Figure 1.2c. After loop transformation

Simple loop transformations lead from Figure 2a to Figure 2b to Figure 2c. Since the nominal feedback system is internally stable, $K(1-PK)^{-1} \in \mathbf{RH}_\infty$. Our baby version of the small gain theorem gives that the system in Figure 2c will be internally stable if

$$\|\Delta PK(1-PK)^{-1}\|_\infty < 1 . \tag{4}$$

In view of (3) a sufficient condition for (4) is

$$\|RK(1-PK)^{-1}\|_\infty \leq 1 . \tag{5}$$

We just used the sub-multiplicative property of the $\mathbf{H}_\infty$-norm:

$$\|FG\|_\infty \leq \|F\|_\infty \|G\|_\infty .$$

We conclude that an $\mathbf{H}_\infty$-norm bound on a weighted closed-loop transfer function, i.e. condition (5), is sufficient for robust stability.

Example 2.

For the second example we need another Hardy space, $\mathbf{H}_2$. It consists of all complex-valued functions $F(s)$ which are analytic in the open right half-plane and satisfy the condition

$$\left[\sup_{\xi > 0} (2\pi)^{-1} \int_{-\infty}^{\infty} |F(\xi + j\omega)|^2 d\omega \right]^{1/2} < \infty .$$

The left-hand side of this inequality is defined to be the $\mathbf{H}_2$-norm of F, $\|F\|_2$. Again, let's focus on real-rational functions. A real-rational function belongs to $\mathbf{RH}_2$ if and only if it's stable and strictly proper. For such a function $F(s)$ it can be proved that its $\mathbf{H}_2$-norm can be obtained by integrating over the imaginary axis:

$$\|F\|_2 = \left[(2\pi)^{-1} \int_{-\infty}^{\infty} |F(j\omega)|^2 d\omega \right]^{1/2} . \tag{6}$$

Consider a one-sided signal $x(t)$ (zero for $t < 0$) and suppose its Laplace transform $\hat{x}(s)$ belongs to $\mathbf{RH}_2$. Then Plancherel's theorem says

$$\int_0^\infty x(t)^2 dt = ||\hat{x}||_2^2 .$$

Thus $||\hat{x}||_2^2$ can be interpreted physically as the energy of the signal $x(t)$.

Next, consider a system with transfer function $F(s)$ in $\mathbf{RH}_\infty$. Let the input and output signals be denoted by $x(t)$ and $y(t)$ respectively. It is easy to see that if $\hat{x} \in \mathbf{RH}_2$ and $||\hat{x}||_2=1$, then $\hat{y} \in \mathbf{RH}_2$ and $||\hat{y}||_2 \le ||F||_\infty$. Thus the $\mathbf{H}_\infty$-norm of the transfer function provides a bound on the system gain

$$\sup\{||\hat{y}||_2 : \hat{x} \in \mathbf{RH}_2, \ ||\hat{x}||_2=1\} .$$

The previous discussion was limited to the familiar class of real-rational functions, but the results are general. In fact the $\mathbf{H}_\infty$-norm of the transfer function equals the system gain. The precise statement is as follows: If $F \in \mathbf{H}_\infty$ and $x \in \mathbf{H}_2$, then $Fx \in \mathbf{H}_2$; moreover

$$||F||_\infty = \sup\{||Fx||_2 : x \in \mathbf{H}_2, ||x||_2=1\} . \tag{7}$$

With these preliminaries let's look at a disturbance attenuation problem. In Figure 1 suppose $v_1=0$ and v_2 represents a disturbance signal referred to the output of the plant P. The objective is to attenuate the effect of v_2 on the output u_2 in a suitably defined sense. As before, we shall assume P and K are real-rational and proper, with at least one of them strictly proper. The transfer function from v_2 to u_2 is the *sensitivity function*

$$S := (1-PK)^{-1} .$$

We shall suppose the disturbance v_2 is not a fixed signal, but can be any function in the class

$$\{v_2 : v_2 = Wx \ \text{for some} \ x \in \mathbf{H}_2, ||x||_2 \le 1\} , \tag{8}$$

where $W, W^{-1} \in \mathbf{H}_\infty$; that is, the disturbance signal class consists of all v_2 in $\mathbf{H}_2$ such that

$$||W^{-1}v_2||_2 \le 1 . \tag{9}$$

Assuming for now that the boundary values $v_2(j\omega)$ and $W(j\omega)$ are well-defined, we can interpret inequality (9) as a constraint on the weighted energy of v_2: the energy-density spectrum $|v_2(j\omega)|^2$ is weighted by the factor $|W(j\omega)|^{-2}$. For example, if $|W(j\omega)|$ were relatively large on a certain frequency band and relatively small off it, then (9) would generate a class of signals having their energy concentrated on that band.

The disturbance attenuation objective can now be stated precisely: minimize the energy of u_2 for the worst v_2 in class (8); equivalently (by virtue of (7)), minimize $\|WS\|_\infty$, the $\mathbf{H}_\infty$-norm of the weighted sensitivity function. In a synthesis problem P and W would be given and K would be chosen to minimize $\|WS\|_\infty$, with the added constraint of internal stability. (In an actual design it may make more sense to employ W as a design parameter, to be adjusted by the designer to shape the magnitude Bode plot of S.)

To recap, we have seen how certain control objectives, robust stability and disturbance attenuation, will be achieved if certain $\mathbf{H}_\infty$-norm bounds are achieved. In Chapter 3 is posed a general $\mathbf{H}_\infty$ optimization problem which includes the above two examples as special cases.

Notes and References

The theory presented in this book was initiated by Zames (1976, 1979, 1981). He formulated the problem of sensitivity reduction by feedback as an optimization problem with an operator norm, in particular, an $\mathbf{H}_\infty$-norm. Relevant contemporaneous works are those of Helton (1976) and Tannenbaum (1977). The important papers of Sarason (1967), and Adamjan, Arov, and Krein (1971) established connections between operator theory and complex function theory, in particular, $\mathbf{H}_\infty$-functions; Helton showed that these two mathematical subjects have useful applications in electrical engineering, namely, in broadband matching. Tannenbaum used (Nevanlinna-Pick) interpolation theory to attack the problem of stabilizing a plant with an unknown gain.

For a survey of the papers in the field the reader may consult Francis and Doyle (1986).

CHAPTER 2

BACKGROUND MATHEMATICS: FUNCTION SPACES

The purpose of this chapter is to collect some elementary concepts and facts from functional analysis.

2.1 Banach and Hilbert Space

Let **X** be a linear space over the field **C** of complex numbers. A *norm* on **X** is a function $x \to \|x\|$ from **X** to the field **R** of reals having the four properties

(i) $\|x\| \geq 0$,

(ii) $\|x\| = 0$ iff $x = 0$,

(iii) $\|cx\| = |c| \, \|x\|, \quad c \in \mathbf{C}$

(iv) $\|x+y\| \leq \|x\| + \|y\|$.

With such a norm we can talk about convergence in **X**: a sequence $\{x_k\}$ in **X** *converges to* x in **X**, and x is the *limit* of the sequence, if the sequence of real numbers $\{\|x_k - x\|\}$ converges to zero; if such x exists, then the sequence is *convergent.* A sequence $\{x_k\}$ is a *Cauchy sequence* if

$$(\forall \epsilon > 0)(\exists \text{ integer } n) \; i, k > n \implies \|x_i - x_k\| < \epsilon .$$

Intuitively, the elements in a Cauchy sequence eventually cluster around each other, so they are "trying to converge". If every Cauchy sequence in **X** is convergent (that is, if every sequence which is trying to converge actually does converge), then **X** is *complete.* A (complex) *Banach space* is a linear space over **C** which has a norm and which is complete.

A subset **S** of a Banach space **X** is a *subspace* if

$$x, y \in \mathbf{S} \implies x + y \in \mathbf{S}$$

and

$$x \in \mathbf{S},\ c \in \mathbf{C} \Rightarrow cx \in \mathbf{S}\ ,$$

and it is *closed* if every sequence in **S** which converges in **X** has its limit in **S**. (If the dimension of **X** is finite, then every subspace is closed, but in general a subspace need not be closed.)

For the definition of Hilbert space start with **X** a linear space over **C**. An *inner product* on **X** is a function $(x,y) \rightarrow \langle x,y \rangle$ from **X**×**X** to **C** having the four properties

(i) $\langle x,x \rangle$ is real and ≥ 0,

(ii) $\langle x,x \rangle = 0$ iff $x = 0$,

(iii) the function $y \rightarrow \langle x,y \rangle$ from **X** to **C** is linear,

(iv) $\overline{\langle x,y \rangle} = \langle y,x \rangle$.

Such an inner product on **X** induces a norm, namely, $\|x\| := \langle x,x \rangle^{1/2}$. With respect to this norm **X** may or may not be complete. A (complex) *Hilbert space* is a linear space over **C** which has an inner product and which is complete.

Two vectors x, y in a Hilbert space **X** are *orthogonal* if $\langle x,y \rangle = 0$. If **S** is a subset of **X**, then $\mathbf{S}^{\perp}$ denotes the set of all vectors in **X** which are orthogonal to every vector in **S**; $\mathbf{S}^{\perp}$ is a closed subspace for any set **S**. If **S** is a closed subspace, then $\mathbf{S}^{\perp}$ is called its *orthogonal complement*, and we have

$$\mathbf{X} = \mathbf{S}^{\perp} \oplus \mathbf{S}\ .$$

This means that every vector in **X** can be written uniquely as the sum of a vector in $\mathbf{S}^{\perp}$ and a vector in **S**.

We shall see in the next two sections several examples of infinite-dimensional Banach and Hilbert spaces, but first let's recall the familiar finite-dimensional examples of each.

The space $\mathbf{C}^n$ is a Hilbert space under the inner product

$$\langle x,y \rangle = x^* y\ .$$

Here x and y are column vectors and * denotes complex-conjugate transpose. The corresponding norm is $\|x\| = (x^* x)^{1/2}$.

The space $\mathbf{C}^{n \times m}$ consists of all $n \times m$ complex matrices. There are several possible norms for $\mathbf{C}^{n \times m}$; for compatibility with our norm on $\mathbf{C}^n$ we shall take the following. The singular values of A in $\mathbf{C}^{n\times m}$ are the square roots of the eigenvalues of the Hermitian matrix A^*A. We define $\|A\|$ to be the largest singular value.

Exercise 1. Prove that

$$\|A\| = \max\{\|Ax\| : \|x\| = 1\} .$$

2.2 Time-Domain Spaces

Consider a signal $x(t)$ defined for all time, $-\infty < t < \infty$, and taking values in $\mathbf{C}^n$. Thus x is a function

$$(-\infty,\infty) \rightarrow \mathbf{C}^n .$$

Restrict x to be square-(Lebesgue) integrable:

$$\int_{-\infty}^{\infty} \|x(t)\|^2 dt < \infty . \tag{1}$$

The norm in (1) is our previously defined norm on $\mathbf{C}^n$. The set of all such signals is the Lebesgue space $\mathbf{L}_2(-\infty,\infty)$. (To simplify notation we suppress the dependence of this space on the integer n.) This space is a Hilbert space under the inner product

$$<x,y> := \int_{-\infty}^{\infty} x(t)^* y(t) dt .$$

Then the norm of x, denoted $\|x\|_2$, equals the square root of the left-hand side of (1).

The set of all signals in $\mathbf{L}_2(-\infty,\infty)$ which equal zero for almost all $t<0$ is a closed subspace, denoted $\mathbf{L}_2[0,\infty)$. Its orthogonal complement (zero for almost all $t>0$) is denoted $\mathbf{L}_2(-\infty,0]$.

2.3 Frequency-Domain Spaces

Consider a function $x(j\omega)$ which is defined for all frequencies, $-\infty<\omega<\infty$, takes values in $\mathbf{C}^n$, and is square-(Lebesgue) integrable with respect to ω. The space of all such functions is denoted $\mathbf{L}_2$ and is a Hilbert space under the inner product

$$<x,y> := (2\pi)^{-1} \int_{-\infty}^{\infty} x(j\omega)^* y(j\omega) d\omega .$$

The norm on $\mathbf{L}_2$ will be denoted $||x||_2$. The space $\mathbf{RL}_2$, the real-rational functions in $\mathbf{L}_2$, consists of n-vectors each component of which is real-rational, strictly proper, and without poles on the imaginary axis.

Next, $\mathbf{H}_2$ is the space of all functions $x(s)$ which are analytic in Re $s>0$, take values in $\mathbf{C}^n$, and satisfy the uniform square-integrability condition

$$||x||_2 := \left[\sup_{\xi>0} (2\pi)^{-1} \int_{-\infty}^{\infty} ||x(\xi+j\omega)||^2 d\omega \right]^{1/2} < \infty .$$

(We have used the same norm symbol for $\mathbf{L}_2(-\infty,\infty)$, $\mathbf{L}_2$, and $\mathbf{H}_2$. Context determines which is intended.) This makes $\mathbf{H}_2$ a Banach space. Functions in $\mathbf{H}_2$ are not defined *a priori* on the imaginary axis, but we can get there in the limit.

Theorem 1. If $x \in \mathbf{H}_2$, then for almost all ω the limit

$$\tilde{x}(j\omega) := \lim_{\xi\to 0} x(\xi+j\omega)$$

exists and $\tilde{x}$ belongs to $\mathbf{L}_2$. Moreover, the mapping $x \to \tilde{x}$ from $\mathbf{H}_2$ to $\mathbf{L}_2$ is linear, injective, and norm-preserving.

It is customary to identify x in $\mathbf{H}_2$ and its boundary function $\tilde{x}$ in $\mathbf{L}_2$. So henceforth we drop the tilde and regard $\mathbf{H}_2$ as a closed subspace of the Hilbert space $\mathbf{L}_2$. The space $\mathbf{RH}_2$ consists of real-rational n-vectors which are stable and strictly proper.

The orthogonal complement $\mathbf{H}_2^\perp$ of $\mathbf{H}_2$ in $\mathbf{L}_2$ is the space of functions $x(s)$ with the following properties: $x(s)$ is analytic in Re $s<0$; $x(s)$ takes values in

$\mathbf{C}^n$; the supremum

$$\sup_{\xi<0} \int_{-\infty}^{\infty} ||x(\xi+j\omega)||^2 d\omega$$

is finite. Again, we identify functions in $\mathbf{H}_2^{\perp}$ and their boundary functions in $\mathbf{L}_2$.

Now we turn to two Banach spaces. First, an $n \times m$ complex-valued matrix $F(j\omega)$ belongs to the Lebesgue space $\mathbf{L}_\infty$ iff $||F(j\omega)||$ is essentially bounded (bounded except possibly on a set of measure zero). The norm just used for $F(j\omega)$ is the norm on $\mathbf{C}^{n\times m}$ introduced in Section 2.1 (largest singular value). Then the $\mathbf{L}_\infty$-norm of F is defined to be

$$||F||_\infty := \text{ess}\sup_{\omega} ||F(j\omega)|| .$$

This makes $\mathbf{L}_\infty$ a Banach space. It is easily checked that $F \in \mathbf{RL}_\infty$ iff F is real-rational, proper, and without poles on the imaginary axis.

The final space is $\mathbf{H}_\infty$. It consists of functions $F(s)$ which are analytic in Re $s>0$, take values in $\mathbf{C}^{n\times m}$, and are bounded in Re $s>0$ in the sense that

$$\sup\{||F(s)|| : \text{Re } s>0\} < \infty .$$

The left-hand side defines the $\mathbf{H}_\infty$-norm of F. There is an analog of Theorem 1 in which $\mathbf{H}_2$ and $\mathbf{L}_2$ are replaced by $\mathbf{H}_\infty$ and $\mathbf{L}_\infty$ respectively: each function in $\mathbf{H}_\infty$ has a unique boundary function in $\mathbf{L}_\infty$, and the mapping from $\mathbf{H}_\infty$-function to boundary $\mathbf{L}_\infty$-function is linear, injective, and norm-preserving. So henceforth we regard $\mathbf{H}_\infty$ as a closed subspace of the Banach space $\mathbf{L}_\infty$. Finally, $\mathbf{RH}_\infty$ consists of those real-rational matrices which are stable and proper.

Let's recap in the real-rational case:

$\mathbf{RL}_2$: vector-valued, strictly proper, no poles on imaginary axis

$\mathbf{RH}_2$: vector-valued, strictly proper, stable

$\mathbf{RH}_2^{\perp}$: vector-valued, strictly proper, no poles in Re $s<0$

$\mathbf{RL}_\infty$: matrix-valued, proper, no poles on imaginary axis

$\mathbf{RH}_\infty$: matrix-valued, proper, stable.

Exercise 1. In the scalar-valued case prove that $\mathbf{RL}_2$ equals the set of all real-rational functions which are strictly proper and have no poles on the imaginary axis.

2.4 Connections

This section contains statements of two basic theorems relating the spaces just introduced. The first, a combined Plancherel and Paley-Wiener theorem, connects the time-domain Hilbert spaces and the frequency-domain Hilbert spaces. A mapping from one Hilbert space to another is a *Hilbert space isomorphism* if it is a linear surjection which preserves inner products. (Such a mapping is continuous, preserves norms, is injective, and has a continuous inverse.)

Theorem 1. The Fourier transform is a Hilbert space isomorphism from $\mathbf{L}_2(-\infty,\infty)$ onto $\mathbf{L}_2$. It maps $\mathbf{L}_2[0,\infty)$ onto $\mathbf{H}_2$ and $\mathbf{L}_2(-\infty,0]$ onto $\mathbf{H}_2^\perp$.

This important theorem says in particular that $\mathbf{H}_2$ is just the set of Laplace transforms of signals in $\mathbf{L}_2[0,\infty)$, i.e. of signals on $t \geq 0$ of finite energy.

The second theorem connects the Hilbert space $\mathbf{H}_2$ with the Banach spaces $\mathbf{L}_\infty$ and $\mathbf{H}_\infty$. For F in $\mathbf{L}_\infty$ and $\mathbf{X}$ denoting either $\mathbf{L}_2$ or $\mathbf{H}_2$, let $F\mathbf{X}$ denote the space $\{Fx : x \in \mathbf{X}\}$.

Theorem 2. (i) If $F \in \mathbf{L}_\infty$, then $F\mathbf{L}_2 \subset \mathbf{L}_2$ and

$$\|F\|_\infty = \sup\{\|Fx\|_2 : x \in \mathbf{L}_2, \|x\|_2 = 1\}$$

$$= \sup\{\|Fx\|_2 : x \in \mathbf{H}_2, \|x\|_2 = 1\} .$$

(ii) If $F \in \mathbf{H}_\infty$, then $F\mathbf{H}_2 \subset \mathbf{H}_2$ and

$$\|F\|_\infty = \sup\{\|Fx\|_2 : x \in \mathbf{H}_2, \|x\|_2 = 1\} .$$

Exercise 1. Let $F(s) = \dfrac{s-1}{s+1}$. Prove that $F\mathbf{H}_\infty$ is closed in $\mathbf{H}_\infty$.

Exercise 2. Let $F(s) = \frac{s}{s+1}$. Prove that $F\mathbf{H}_2$ is not closed in $\mathbf{H}_2$.

Notes and References

For Banach and Hilbert spaces the reader is referred to any good book on functional analysis, for instance Conway (1985). Standard references for Hardy spaces are Duren (1970), Garnett (1981), Hoffman (1962), and Rudin (1966). These references deal mainly with scalar-valued functions. For vector- and operator-valued functions, see Sz.-Nagy and Foias (1970) and Rosenblum and Rovnyak (1985). Theorem 3.1 can be derived from the results in Chapter 11 in Duren (1970). Theorem 4.1 is in Paley and Wiener (1934). For an alternative proof see Dym and McKean (1972). A complete proof of Theorem 4.2 is not readily available in the literature; a starting point could be Problem 64 in Halmos (1982), Theorem II.6.7 in Desoer and Vidyasagar (1975), or Theorem II.1.5 in Conway (1985).

CHAPTER 3

THE STANDARD PROBLEM

The standard set-up is shown in Figure 1. In this figure w, u, z, and y are vector-valued signals: w is the exogenous input, typically consisting of command signals, disturbances, and sensor noises; u is the control signal; z is the output to be controlled, its components typically being tracking errors, filtered actuator signals, etc.; and y is the measured output. The transfer matrices G and K are, by assumption, real-rational and proper: G represents a generalized plant, the fixed part of the system, and K represents a controller. Partition G as

$$G = \begin{bmatrix} G_{11} & G_{12} \\ G_{21} & G_{22} \end{bmatrix}.$$

Then Figure 1 stands for the algebraic equations

$$z = G_{11}w + G_{12}u$$

$$y = G_{21}w + G_{22}u$$

$$u = Ky.$$

To define what it means for K to stabilize G, introduce two additional inputs, v_1 and v_2, as in Figure 2. The equation relating the three inputs w, v_1, v_2 and the three signals z, u, y is

$$\begin{bmatrix} I & -G_{12} & 0 \\ 0 & I & -K \\ 0 & -G_{22} & I \end{bmatrix} \begin{bmatrix} z \\ u \\ y \end{bmatrix} = \begin{bmatrix} G_{11} & 0 & 0 \\ 0 & I & 0 \\ G_{21} & 0 & I \end{bmatrix} \begin{bmatrix} w \\ v_1 \\ v_2 \end{bmatrix}.$$

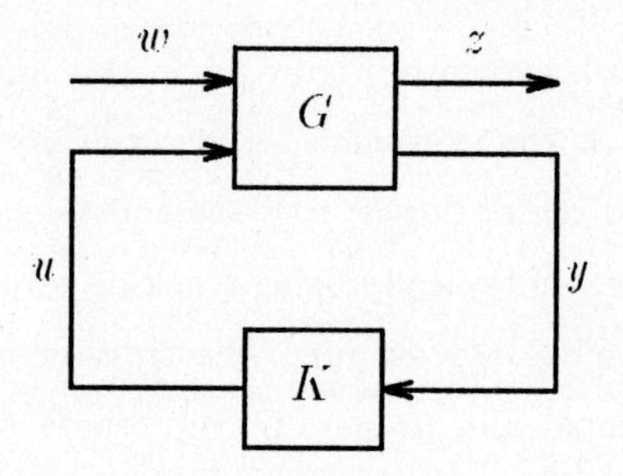

Figure 3.1. The standard block diagram

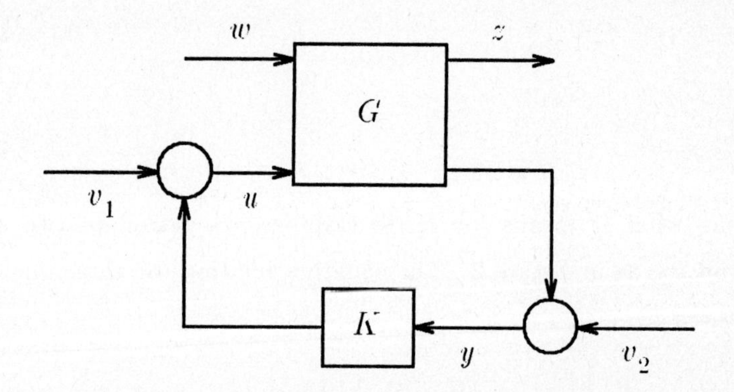

Figure 3.2. Diagram for stability definition

It simplifies the theory to guarantee that the proper real-rational matrix

$$\begin{bmatrix} I & -G_{12} & 0 \\ 0 & I & -K \\ 0 & -G_{22} & I \end{bmatrix}$$

has a proper real-rational inverse for every proper real-rational K. A simple sufficient condition for this is that G_{22} be strictly proper. Accordingly, this will be *assumed* hereafter. Then the nine transfer matrices from w, v_1, v_2 to z, u, y are proper. If they are stable, i.e. they belong to $\mathbf{RH}_\infty$, then we say that K *stabilizes* G. This is the usual notion of internal stability. An equivalent definition in terms of state-space models is as follows. Take minimal state-space realizations of G and K and in Figure 1 set the input w to zero. Then K stabilizes G if and only if the state vectors of G and K tend to zero from every initial condition.

The *standard problem* is this: find a real-rational proper K to minimize the $\mathbf{H}_\infty$-norm of the transfer matrix from w to z under the constraint that K stabilize G. The transfer matrix from w to z is a linear-fractional transformation of K:

$$z = [G_{11} + G_{12}K(I-G_{22}K)^{-1}G_{21}]w.$$

Following are three examples of the standard problem.

A Model-Matching Problem

In Figure 3 the transfer matrix T_1 represents a "model" which is to be matched by the cascade $T_2 Q T_3$ of three transfer matrices T_2, T_3, and Q. Here, T_i $(i=1\text{–}3)$ are given and the "controller" Q is to be designed. It is assumed that $T_i \in \mathbf{RH}_\infty$ $(i=1\text{–}3)$ and it is required that $Q \in \mathbf{RH}_\infty$. Thus the four blocks in Figure 3 represent stable linear systems.

For our purposes the *model-matching criterion* is

$$\sup \{\|z\|_2 : w \in \mathbf{H}_2, \|w\|_2 \le 1\} = \text{minimum}.$$

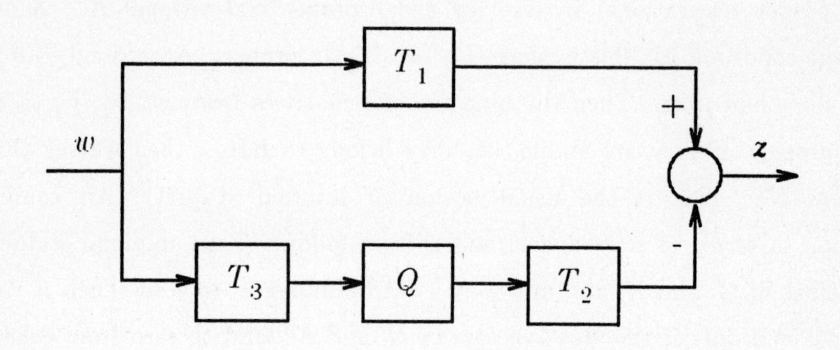

Figure 3.3. Model-matching

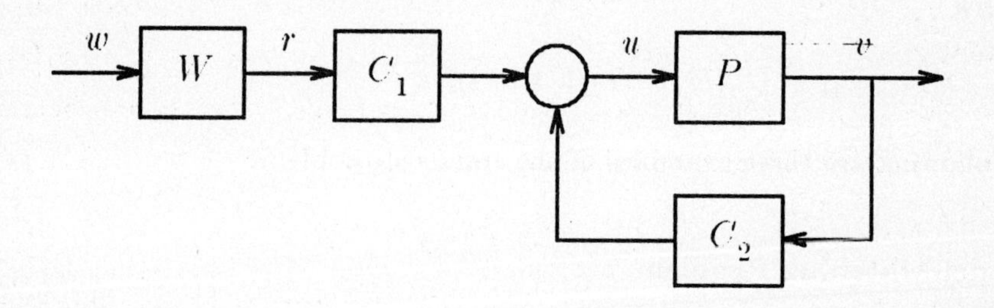

Figure 3.4. Tracking

Thus the energy of the error z is to be minimized for the worst input w of unit energy. In view of Theorem 2.4.2 an equivalent criterion is

$$\|T_1 - T_2 Q T_3\|_\infty = \text{minimum}.$$

This model-matching problem can be recast as a standard problem by defining

$$G := \begin{bmatrix} T_1 & T_2 \\ T_3 & 0 \end{bmatrix}$$

$$K := -Q,$$

so that Figure 3 becomes equivalent to Figure 1. The constraint that K stabilize G is then equivalent to the constraint that $Q \in \mathbf{RH}_\infty$.

This version of the model-matching problem is not very important *per se* ; its significance in the context of this course arises from the fact that the standard problem can be transformed to the model-matching problem (Chapter 4), which is considerably simpler.

A Tracking Problem

Figure 4 shows a plant P whose output, v, is to track a reference signal r. The plant input, u, is generated by passing r and v through controllers C_1 and C_2 respectively. It is postulated that r is not a known fixed signal, but, as in Chapter 1, may be modeled as belonging to the class

$$\{ r : r = Ww \text{ for some } w \in \mathbf{H}_2, \|w\|_2 \leq 1 \}.$$

Here P and W are given and C_1 and C_2 are to be designed. These four transfer matrices are assumed to be real-rational and proper.

The tracking error signal is $r - v$. Let's take the cost function to be

$$(\|r - v\|_2^2 + \|\rho u\|_2^2)^{1/2}, \tag{1}$$

where ρ is a positive scalar weighting factor. The reason for including ρu in (1) is to ensure the existence of an optimal proper controller; for $\rho = 0$ "optimal" controllers tend to be improper. Note that (1) equals the $\mathbf{H}_2$-norm of

$$z := \begin{bmatrix} r - v \\ \rho u \end{bmatrix}.$$

Thus the *tracking criterion* is taken to be

$$\sup\{\|z\|_2 : w \in \mathbf{H}_2, \|w\|_2 \le 1\} = \text{minimum}.$$

The equivalent standard problem is obtained by defining

$$y := \begin{bmatrix} r \\ v \end{bmatrix}, \quad K := [C_1 \quad C_2]$$

$$G := \begin{bmatrix} G_{11} & G_{12} \\ G_{21} & G_{22} \end{bmatrix}$$

$$G_{11} := \begin{bmatrix} W \\ 0 \end{bmatrix}, \quad G_{12} := \begin{bmatrix} -P \\ \rho I \end{bmatrix}$$

$$G_{21} := \begin{bmatrix} W \\ 0 \end{bmatrix}, \quad G_{22} := \begin{bmatrix} 0 \\ P \end{bmatrix}.$$

A Robust Stabilization Problem

This example has already been discussed in Chapter 1. The system under consideration is shown in Figure 1.2a. Assume P is a strictly proper nominal plant and let R be a scalar-valued (radius) function in $\mathbf{RH}_\infty$. Now define a family $\mathbf{P}$ of neighbouring plants to consist of all strictly proper real-rational matrices $P+\Delta P$ having the same number (in terms of McMillan degree) of poles in $\text{Re } s \geq 0$ as has P, where the perturbation ΔP satisfies the bound

$$\|\Delta P(j\omega)\| < |R(j\omega)| \quad \text{for all } 0 \leq \omega \leq \infty .$$

For a real-rational proper K the *robust stability criterion* is that K stabilize all plants in $\mathbf{P}$. Stability means internal stability, that the four transfer matrices in Figure 1.2a from v_1, v_2 to u, y all belong to $\mathbf{RH}_\infty$.

We saw in Chapter 1 that robust stability is guaranteed by a small gain condition.

Lemma 1. A real-rational proper K stabilizes all plants in $\mathbf{P}$ iff K stabilizes the nominal plant P and

$$\|RK(I-PK)^{-1}\|_\infty \leq 1.$$

We can convert to the set-up of the standard problem by defining G so that in Figure 1 the transfer matrix from w to z equals $RK(I-PK)^{-1}$. This is accomplished by

$$G := \begin{bmatrix} 0 & RI \\ I & P \end{bmatrix}.$$

Then Lemma 1 implies that the following two conditions are equivalent: K achieves robust stability for the original system (Figure 1.2a); in Figure 1 K stabilizes G and makes the transfer matrix from w to z have $\mathbf{H}_\infty$-norm ≤ 1.

Notes and References

The standard problem as posed in this chapter is based on Doyle (1984). For treatments of the tracking example see Vidyasagar (1985b) and Wang and Pearson (1984), and for robust stabilization see, for example, Kimura (1984). Lemma 1 is due to Doyle and Stein (1981) and Chen and Desoer (1982).

There are several other examples of the standard problem, for example, the weighted sensitivity problem (Zames (1981)) and the mixed sensitivity problem (Verma and Jonckheere (1984), Kwakernaak (1985)).

CHAPTER 4

STABILITY THEORY

In this chapter it is shown how the standard problem can be reduced to the model-matching problem. The procedure is to parametrize, via a parameter matrix Q in $\mathbf{RH}_\infty$, all K's which stabilize G.

4.1 Coprime Factorization Over $\mathbf{RH}_\infty$

Recall that two polynomials $f(s)$ and $g(s)$, with, say, real coefficients, are said to be *coprime* if their greatest common divisor is 1 (equivalently, they have no common zeros). It follows from Euclid's algorithm that f and g are coprime iff there exist polynomials $x(s)$ and $y(s)$ such that

$$fx + gy = 1 . \tag{1}$$

Such an equation is called a Bezout identity.

We are going to take the practical route and define two functions f and g in $\mathbf{RH}_\infty$ to be *coprime* (over $\mathbf{RH}_\infty$) if there exist x, y in $\mathbf{RH}_\infty$ such that (1) holds. (The more primitive, but equivalent, definition is that f and g are coprime if every common divisor of f and g is invertible in $\mathbf{RH}_\infty$, i.e.

$$h,\ fh^{-1},\ gh^{-1} \in \mathbf{RH}_\infty \Rightarrow h^{-1} \in \mathbf{RH}_\infty .)$$

More generally, two matrices F and G in $\mathbf{RH}_\infty$ are *right-coprime* (over $\mathbf{RH}_\infty$) if they have equal number of columns and there exist matrices X and Y in $\mathbf{RH}_\infty$ such that

$$[X \quad Y] \begin{bmatrix} F \\ G \end{bmatrix} = XF + YG = I .$$

This is equivalent to saying that the matrix $\begin{bmatrix} F \\ G \end{bmatrix}$ is left-invertible in $\mathbf{RH}_\infty$.

Similarly, two matrices F and G in $\mathbf{RH}_\infty$ are *left-coprime* (over $\mathbf{RH}_\infty$) if they have equal number of rows and there exist X and Y in $\mathbf{RH}_\infty$ such that

$$[F \quad G]\begin{bmatrix} X \\ Y \end{bmatrix} = FX + GY = I \ ;$$

equivalently, $[F \quad G]$ is right-invertible in $\mathbf{RH}_\infty$.

Now let G be a proper real-rational matrix. A *right-coprime factorization* of G is a factorization $G = NM^{-1}$ where N and M are right-coprime $\mathbf{RH}_\infty$-matrices. Similarly, a *left-coprime factorization* has the form $G = \tilde{M}^{-1}\tilde{N}$ where $\tilde{N}$ and $\tilde{M}$ are left-coprime. Of course implicit in these definitions is the requirement that M and $\tilde{M}$ be square and nonsingular. We shall require special coprime factorizations, as described in the next lemma.

Lemma 1. For each proper real-rational matrix G there exist eight $\mathbf{RH}_\infty$-matrices satisfying the equations

$$G = NM^{-1} = \tilde{M}^{-1}\tilde{N} \tag{2}$$

$$\begin{bmatrix} \tilde{X} & -\tilde{Y} \\ -\tilde{N} & \tilde{M} \end{bmatrix} \begin{bmatrix} M & Y \\ N & X \end{bmatrix} = I \ . \tag{3}$$

Equations (2) and (3) together constitute a *doubly-coprime factorization* of G. It should be apparent that N and M are right-coprime and $\tilde{N}$ and $\tilde{M}$ are left-coprime; for example, (3) implies

$$[\tilde{X} \quad -\tilde{Y}]\begin{bmatrix} M \\ N \end{bmatrix} = I \ ,$$

proving right-coprimeness.

It's useful to prove Lemma 1 constructively by deriving explicit formulas for the eight matrices. The formulas use state-space realizations, and hence are readily amenable to computer implementation.

We start with a state-space realization of G,

$$G(s) = D + C(s - A)^{-1}B \tag{4}$$

A, B, C, D real matrices,

with (A, B) stabilizable and (C, A) detectable. It's convenient to introduce a new data structure: let

$$[A, B, C, D]$$

stand for the transfer matrix

$$D + C(s-A)^{-1}B .$$

Now introduce state, input, and output vectors x, u, and y respectively so that $y = Gu$ and

$$\dot{x} = Ax + Bu \tag{5a}$$

$$y = Cx + Du . \tag{5b}$$

Next, choose a real matrix F such that $A_F := A + BF$ is stable (all eigenvalues in Re $s < 0$) and define the vector $v := u - Fx$ and the matrix $C_F := C + DF$. Then from (5) we get

$$\dot{x} = A_F x + Bv$$

$$u = Fx + v$$

$$y = C_F x + Dv .$$

Evidently from these equations the transfer matrix from v to u is

$$M(s) := [A_F, B, F, I] \tag{6a}$$

and that from v to y is

$$N(s) := [A_F, B, C_F, D] . \tag{6b}$$

Therefore

$$u = Mv , \quad y = Nv$$

so that $y = NM^{-1}u$, i.e. $G = NM^{-1}$.

Similarly, by choosing a real matrix H so that $A_H := A + HC$ is stable and defining

$$B_H := B + HD$$

$$\tilde{M}(s) := [A_H, H, C, I] \tag{6c}$$

$$\tilde{N}(s) := [A_H, B_H, C, D], \tag{6d}$$

we get $G = \tilde{M}^{-1}\tilde{N}$. (This can be derived as above by starting with $G(s)^T$ instead of $G(s)$.) Thus we've obtained four matrices in $\mathbf{RH}_\infty$ satisfying (2).

Formulas for the other four matrices to satisfy (3) are as follows:

$$X(s) := [A_F, -H, C_F, I] \tag{7a}$$

$$Y(s) := [A_F, -H, F, 0] \tag{7b}$$

$$\tilde{X}(s) := [A_H, -B_H, F, I] \tag{7c}$$

$$\tilde{Y}(s) := [A_H, -H, F, 0]. \tag{7d}$$

The explanation of where these latter four formulas come from is deferred to Section 4.

Exercise 1. Verify that the matrices in (6) and (7) satisfy (3).

Example 1.

As an illustration of the use of these formulas, consider the scalar-valued example

$$G(s) = \frac{s-1}{s(s-2)}.$$

A minimal realization is

$$G(s) = [A, B, C, D]$$

$$A = \begin{bmatrix} 0 & 1 \\ 0 & 2 \end{bmatrix}, \quad B = \begin{bmatrix} 0 \\ 1 \end{bmatrix}$$

$$C = [-1 \ \ 1], \quad D = 0.$$

Choosing F to place the eigenvalues of A_F (arbitrarily) at $\{-1, -1\}$, we get

$$F = [-1 \ \ -4]$$

$$A_F = \begin{bmatrix} 0 & 1 \\ -1 & -2 \end{bmatrix}.$$

Then

$$N(s) = [A_F, B, C, 0]$$

$$= \frac{s-1}{(s+1)^2}$$

and

$$M(s) = [A_F, B, F, 1]$$

$$= \frac{s(s-2)}{(s+1)^2}.$$

Similarly, the assignment

$$H = \begin{bmatrix} -5 \\ -9 \end{bmatrix}$$

yields

$$A_H = \begin{bmatrix} 5 & -4 \\ 9 & -7 \end{bmatrix}$$

$$X(s) = [A_F, -H, C, 1]$$

$$= \frac{s^2+6s-23}{(s+1)^2}$$

$$Y(s) = [A_F, -H, F, 0]$$

$$= \frac{-41s+1}{(s+1)^2}.$$

Finally, in this example we have

$$\tilde{N} = N, \ \tilde{M} = M, \ \tilde{X} = X, \ \tilde{Y} = Y .$$

4.2 Stability

This section provides a test for when a proper real-rational K stabilizes G. Introduce left- and right-coprime factorizations

$$G = NM^{-1} = \tilde{M}^{-1}\tilde{N} \tag{1a}$$

$$K = UV^{-1} = \tilde{V}^{-1}\tilde{U} . \tag{1b}$$

Theorem 1. The following are equivalent statements about K:

(i) K stabilizes G,

(ii) $$\begin{bmatrix} M & \begin{bmatrix} 0 \\ I \end{bmatrix} U \\ [0 \;\; I] N & V \end{bmatrix}^{-1} \in \mathbf{RH}_\infty ,$$

(iii) $$\begin{bmatrix} \tilde{M} & \tilde{N} \begin{bmatrix} 0 \\ I \end{bmatrix} \\ \tilde{U} [0 \;\; I] & \tilde{V} \end{bmatrix}^{-1} \in \mathbf{RH}_\infty .$$

The idea underlying the equivalence of (i) and (ii) is simply that the determinant of the matrix in (ii) is the least common denominator (in $\mathbf{RH}_\infty$) of all the transfer functions from w, v_1, v_2 to z, u, y; hence the determinant must be invertible for all these transfer functions to belong to $\mathbf{RH}_\infty$, and conversely.

The proof of Theorem 1 requires a preliminary result. Insert the factorizations (1) into Figure 1, split apart the factors, and introduce two new signals ξ and η to get Figure 2.

Lemma 1. The nine transfer matrices in Figure 1 from w, v_1, v_2 to z, u, y belong to $\mathbf{RH}_\infty$ iff the six transfer matrices in Figure 2 from w, v_1, v_2 to ξ, η belong to $\mathbf{RH}_\infty$.

Proof. (If) This direction follows immediately from the equations

$$\begin{bmatrix} z \\ y \end{bmatrix} = N \xi + \begin{bmatrix} 0 \\ v_2 \end{bmatrix}$$

$$u = U \eta + v_1 ,$$

which in turn follow from Figure 2.

(Only if) By right-coprimeness there exist $\mathbf{RH}_\infty$-matrices X and Y such that

$$XM + YN = I .$$

Hence

$$\xi = XM \xi + YN \xi . \tag{2}$$

But from Figure 2

$$M\xi = \begin{bmatrix} w \\ u \end{bmatrix}, \quad N\xi = \begin{bmatrix} z \\ y - v_2 \end{bmatrix}.$$

Substitution into (2) gives

$$\xi = X \begin{bmatrix} 0 \\ u \end{bmatrix} + Y \begin{bmatrix} z \\ y \end{bmatrix} + X \begin{bmatrix} w \\ 0 \end{bmatrix} - Y \begin{bmatrix} 0 \\ v_2 \end{bmatrix}.$$

Hence the three transfer matrices from w, v_1, v_2 to ξ belong to $\mathbf{RH}_\infty$.

A similar argument works for the remaining three transfer matrices to η. □

Proof of Theorem 1. We shall prove the equivalence of (i) and (ii). First, let's see that the matrix displayed in (ii) is indeed nonsingular, i.e. its inverse exists as a rational matrix. We have

$$\begin{bmatrix} M & \begin{bmatrix} 0 \\ I \end{bmatrix} U \\ [0 \;\; I]N & V \end{bmatrix} = \begin{bmatrix} I & \begin{bmatrix} 0 \\ I \end{bmatrix} K \\ [0 \;\; I]G & I \end{bmatrix} \begin{bmatrix} M & 0 \\ 0 & V \end{bmatrix}$$

$$= \begin{bmatrix} I & 0 & 0 \\ 0 & I & K \\ G_{21} & G_{22} & I \end{bmatrix} \begin{bmatrix} M & 0 \\ 0 & V \end{bmatrix}. \tag{3}$$

Now

$$\begin{bmatrix} M & 0 \\ 0 & V \end{bmatrix}$$

is nonsingular because both M and V are. Also, since G_{22} is strictly proper, we have that

$$\begin{bmatrix} I & 0 & 0 \\ 0 & I & K \\ G_{21} & G_{22} & I \end{bmatrix}$$

is nonsingular when evaluated at $s = \infty$: its determinant equals 1 at $s = \infty$. Thus both matrices on the right-hand side of (3) are nonsingular.

The equations corresponding to Figure 2 are

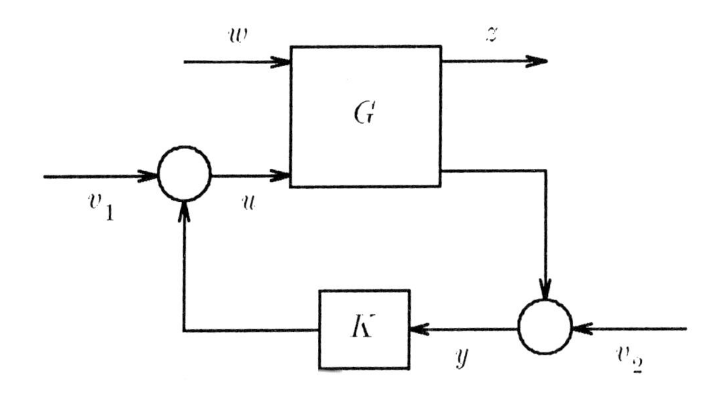

Figure 4.2.1. Diagram for stability definition

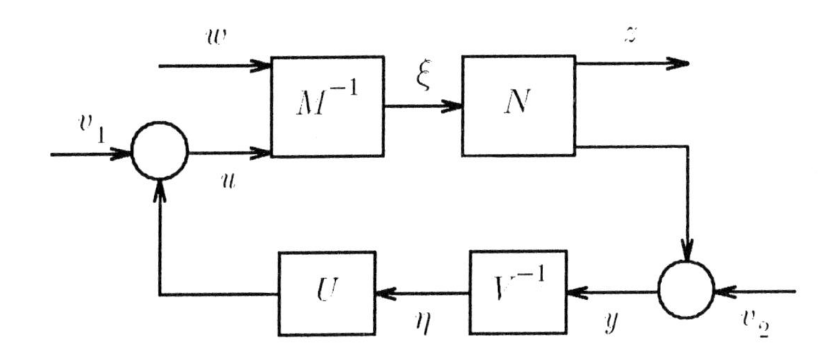

Figure 4.2.2. With internal signals

$$\begin{bmatrix} M & -\begin{bmatrix} 0 \\ I \end{bmatrix} U \\ -[0 \ \ I\,]N & V \end{bmatrix} \begin{bmatrix} \xi \\ \eta \end{bmatrix} = \begin{bmatrix} \begin{bmatrix} w \\ v_1 \end{bmatrix} \\ v_2 \end{bmatrix} .$$

Thus by Lemma 1 K stabilizes G iff

$$\begin{bmatrix} M & -\begin{bmatrix} 0 \\ I \end{bmatrix} U \\ -[0 \ \ I\,]N & V \end{bmatrix}^{-1} \in \mathbf{RH}_\infty .$$

But this is equivalent to (ii). □

Exercise 1. Prove equivalence of (i) and (iii) in Theorem 1.

4.3 Stabilizability

Let's say that G is *stabilizable* if there exists a (proper real-rational) K which stabilizes it. Not every G is stabilizable; an obvious non-stabilizable G is $G_{12}=0$, $G_{21}=0$, $G_{22}=0$, G_{11} unstable. In this example, the unstable part of G is disconnected from u and y. In terms of a state-space model G is stabilizable iff its unstable modes are controllable from u (stabilizability) and observable from y (detectability). The next result is a stabilizability test in terms of left- and right-coprime factorizations

$$G = NM^{-1} = \tilde{M}^{-1}\tilde{N} .$$

Theorem 1. The following conditions are equivalent:

(i) G is stabilizable,

(ii) M, $[0 \ \ I\,]N$ are right-coprime and

M, $\begin{bmatrix} 0 \\ I \end{bmatrix}$ are left-coprime,

(iii) $\tilde{M}$, $\tilde{N}\begin{bmatrix} 0 \\ I \end{bmatrix}$ are left-coprime and

$\tilde{M}$, $[0 \ \ I\,]$ are right-coprime.

The proof requires some preliminaries. The reader will recall the following fact. For each real matrix F there exist real matrices G and H such that

$$F = G \begin{bmatrix} I & 0 \\ 0 & 0 \end{bmatrix} H.$$

The matrices G and H may be obtained by elementary row and column operations, and the size of the identity matrix equals the rank of F. The following analogous result for $\mathbf{RH}_\infty$-matrices is stated without proof.

Lemma 1. For each matrix F in $\mathbf{RH}_\infty$ there exist matrices G, H, and F_1 in $\mathbf{RH}_\infty$ satisfying the equation

$$F = G \begin{bmatrix} F_1 & 0 \\ 0 & 0 \end{bmatrix} H$$

and having the properties that G and H are invertible in $\mathbf{RH}_\infty$ and F_1 is nonsingular.

This result is now used to prove the following useful fact that if M and N are right-coprime, then the matrix $\begin{bmatrix} M \\ N \end{bmatrix}$ can be filled out to yield a square matrix which is invertible in $\mathbf{RH}_\infty$.

Lemma 2. Let M and N be $\mathbf{RH}_\infty$-matrices with equal number of columns. Then M and N are right-coprime iff there exist matrices U and V in $\mathbf{RH}_\infty$ such that

$$\begin{bmatrix} M & U \\ N & V \end{bmatrix}^{-1} \in \mathbf{RH}_\infty .$$

Proof. (If) Define

$$\begin{bmatrix} X & Y \\ ? & ? \end{bmatrix} := \begin{bmatrix} M & U \\ N & V \end{bmatrix}^{-1},$$

where a question mark denotes an irrelevant block. Then

$$[X \quad Y]\begin{bmatrix} M \\ N \end{bmatrix} = I \ ,$$

so M and N are right-coprime.

(Only if) Define

$$F := \begin{bmatrix} M \\ N \end{bmatrix}$$

and bring in matrices G , H , and F_1 as per Lemma 1. Since F is left-invertible in $\mathbf{RH}_\infty$ (by right-coprimeness), it follows that

$$\begin{bmatrix} F_1 & 0 \\ 0 & 0 \end{bmatrix}$$

is left-invertible in $\mathbf{RH}_\infty$ too. But then it must have the form

$$\begin{bmatrix} F_1 \\ 0 \end{bmatrix}$$

with $F_1^{-1} \in \mathbf{RH}_\infty$. Defining

$$K := G\begin{bmatrix} F_1 H & 0 \\ 0 & I \end{bmatrix},$$

we get

$$\begin{bmatrix} M \\ N \end{bmatrix} = K\begin{bmatrix} I \\ 0 \end{bmatrix}.$$

Thus the definition

$$\begin{bmatrix} U \\ V \end{bmatrix} := K\begin{bmatrix} 0 \\ I \end{bmatrix}$$

gives the desired result, that

$$\begin{bmatrix} M & U \\ N & V \end{bmatrix} = K$$

is invertible in $\mathbf{RH}_\infty$. □

The obvious dual of Lemma 2 is that M and N are left-coprime iff there exist U and V such that

$$\begin{bmatrix} M & N \\ U & V \end{bmatrix}^{-1} \in \mathbf{RH}_\infty .$$

Proof of Theorem 1. We shall prove equivalence of (i) and (ii).

(i) $\Rightarrow$ (ii): If G is stabilizable, then by Theorem 2.1 there exist U and V in $\mathbf{RH}_\infty$ such that

$$\begin{bmatrix} M & \begin{bmatrix} 0 \\ I \end{bmatrix} U \\ [0 \ \ I]N & V \end{bmatrix}^{-1} \in \mathbf{RH}_\infty .$$

This implies by Lemma 2 and its dual that

$$M,\ [0 \ \ I]N \text{ are right-coprime}$$

and

$$M,\ \begin{bmatrix} 0 \\ I \end{bmatrix} U \text{ are left-coprime.}$$

But the latter condition implies left-coprimeness of

$$M,\ \begin{bmatrix} 0 \\ I \end{bmatrix} .$$

(ii) $\Rightarrow$ (i): Choose, by right-coprimeness and Lemma 2, matrices X and Y in $\mathbf{RH}_\infty$ such that

$$\begin{bmatrix} M & X \\ [0 \ \ I]N & Y \end{bmatrix}^{-1} \in \mathbf{RH}_\infty .$$

Also, choose, by left-coprimeness, matrices R and T in $\mathbf{RH}_\infty$ such that

$$\begin{bmatrix} M & \begin{bmatrix} 0 \\ I \end{bmatrix} \end{bmatrix} \begin{bmatrix} R \\ T \end{bmatrix} = I . \tag{1}$$

Now define

$$U := TX \tag{2a}$$

$$V := Y - [0\ I]NRX\ . \tag{2b}$$

Then we have from (1) and (2) that

$$\begin{bmatrix} M & X \\ [0\ I]N & Y \end{bmatrix} \begin{bmatrix} I & -RX \\ 0 & I \end{bmatrix}$$

$$= \begin{bmatrix} M & \begin{bmatrix} 0 \\ I \end{bmatrix} U \\ [0\ I]N & V \end{bmatrix} . \tag{3}$$

The two matrices on the left in (3) have inverses in $\mathbf{RH}_\infty$, hence so does the matrix on the right in (3).

The next step is to show that V is nonsingular. We have

$$\begin{bmatrix} M & \begin{bmatrix} 0 \\ I \end{bmatrix} U \\ [0\ I]N & V \end{bmatrix} = \begin{bmatrix} I & \begin{bmatrix} 0 \\ I \end{bmatrix} U \\ [0\ I]G & V \end{bmatrix} \begin{bmatrix} M & 0 \\ 0 & I \end{bmatrix}$$

$$= \begin{bmatrix} I & 0 & 0 \\ 0 & I & U \\ G_{21} & G_{22} & V \end{bmatrix} \begin{bmatrix} M & 0 \\ 0 & I \end{bmatrix} . \tag{4}$$

Evaluate all the matrices in (4) at $s = \infty$; then take determinants of both sides noting that G_{22} is strictly proper and the matrix on the left-hand side of (4) is invertible in $\mathbf{RH}_\infty$. This gives

$$0 \neq \det V(\infty) \det M(\infty) .$$

Thus $\det V(\infty) \neq 0$, i.e. V^{-1} exists. Hence we can define $K := UV^{-1}$.

Next, note that U and V are right-coprime (this follows from invertibility in $\mathbf{RH}_\infty$ of the matrix on the right-hand side of (3)). We conclude from Theorem 2.1 that K stabilizes G. □

Exercise 1. Prove equivalence of (i) and (iii) in Theorem 1.

Hereafter, G will be assumed to be stabilizable. Intuitively, this implies that G and G_{22} share the same unstable poles (counting multiplicities), so to stabilize G it is enough to stabilize G_{22}. Let's define the latter concept explicitly: K *stabilizes* G_{22} if in Figure 2.1 the four transfer matrices from v_1 and v_2 to u and y belong to $\mathbf{RH}_\infty$.

Theorem 2. K stabilizes G iff K stabilizes G_{22}.

The necessity part of the theorem follows from the definitions. To prove sufficiency we need a result analogous to Lemma 2.1.

Lemma 3. The four transfer matrices in Figure 2.1 from v_1, v_2 to u, y belong to $\mathbf{RH}_\infty$ iff the four transfer matrices in Figure 2.2 from v_1, v_2 to ξ, η belong to $\mathbf{RH}_\infty$.

The proof is omitted, it being entirely analogous to that of Lemma 2.1.

Proof of Theorem 2. Suppose K stabilizes G_{22}. To prove that K stabilizes G it suffices to show, by Lemma 2.1, that the six transfer matrices in Figure 2.2 from w, v_1, v_2 to ξ, η belong to $\mathbf{RH}_\infty$. But by Lemma 3 we know that those from v_1, v_2 to ξ, η do. So it remains to show that the two from w to ξ, η belong to $\mathbf{RH}_\infty$.

Set $v_1=0$ and $v_2=0$ in Figure 2.2 and write the corresponding equations:

$$M\xi = \begin{bmatrix} I \\ 0 \end{bmatrix} w + \begin{bmatrix} 0 \\ I \end{bmatrix} U\eta \tag{5}$$

$$V\eta = [0 \quad I] N\xi \,. \tag{6}$$

By left-coprimeness there exist matrices R and T in $\mathbf{RH}_\infty$ such that

$$\begin{bmatrix} M & \begin{bmatrix} 0 \\ I \end{bmatrix} \end{bmatrix} \begin{bmatrix} R \\ T \end{bmatrix} = I \,. \tag{7}$$

Post-multiply (7) by $\begin{bmatrix} I \\ 0 \end{bmatrix} w$ to get

$$MR \begin{bmatrix} I \\ 0 \end{bmatrix} w + \begin{bmatrix} 0 \\ I \end{bmatrix} T \begin{bmatrix} I \\ 0 \end{bmatrix} w = \begin{bmatrix} I \\ 0 \end{bmatrix} w \,. \tag{8}$$

Now subtract (8) from (5), rearrange, and define

$$\xi_1 := \xi - R \begin{bmatrix} I \\ 0 \end{bmatrix} w \tag{9}$$

$$v_1 := T \begin{bmatrix} I \\ 0 \end{bmatrix} w \tag{10}$$

to get

$$M \xi_1 = \begin{bmatrix} 0 \\ I \end{bmatrix} (v_1 + U \eta) . \tag{11}$$

Also, rearrange (6) and define

$$v_2 := [0 \;\; I] NR \begin{bmatrix} I \\ 0 \end{bmatrix} w \tag{12}$$

to get

$$V \eta = [0 \;\; I] N \xi_1 + v_2 . \tag{13}$$

The block diagram corresponding to (11) and (13) is Figure 1. By Lemma 3 and the fact that K stabilizes G_{22} we know that the transfer matrices in Figure 1 from v_1, v_2 to ξ_1, η belong to $\mathbf{RH}_\infty$. But by (10) and (12) those from w to v_1, v_2 belong to $\mathbf{RH}_\infty$. Hence those from w to ξ_1, η belong to $\mathbf{RH}_\infty$. Finally, we conclude from (9) that the transfer matrix from w to ξ belongs to $\mathbf{RH}_\infty$. □

Exercise 2. Suppose $G_{11} = G_{12} = G_{21} = G_{22}$. Prove that G is stabilizable.

4.4 Parametrization

This section contains a parametrization of all K's which stabilize G_{22}. To simplify notation slightly, in this section the subscripts 22 on G_{22} are dropped. The relevant block diagram is Figure 1.

Bring in a doubly-coprime factorization of G,

$$G = NM^{-1} = \tilde{M}^{-1}\tilde{N}$$

$$\begin{bmatrix} \tilde{X} & -\tilde{Y} \\ -\tilde{N} & \tilde{M} \end{bmatrix} \begin{bmatrix} M & Y \\ N & X \end{bmatrix} = I , \tag{1}$$

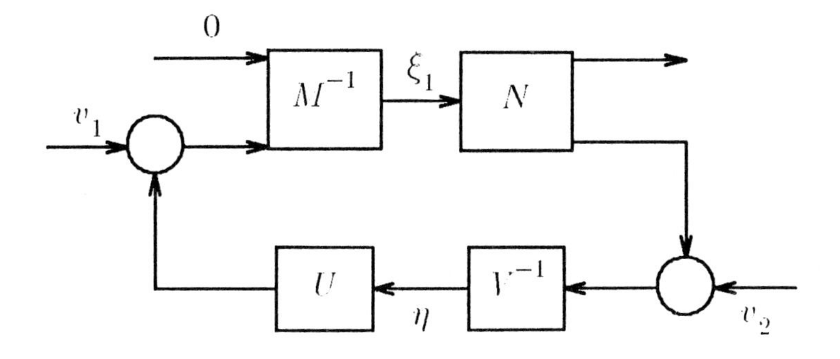

Figure 4.3.1. For proof of Theorem 4.3.2

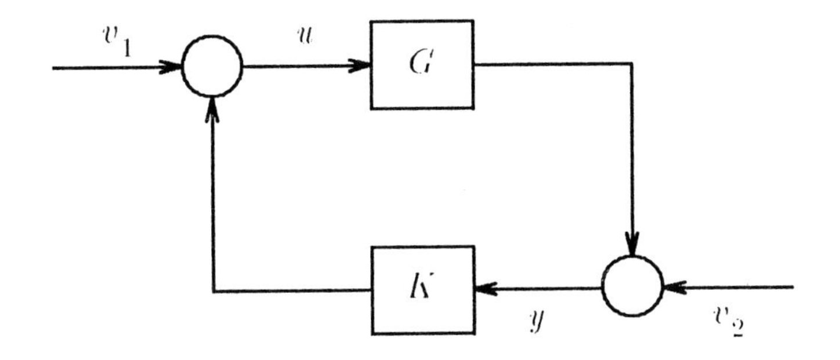

Figure 4.4.1. Diagram for controller parametrization

and coprime factorizations (not necessarily doubly-coprime) of K,

$$K = UV^{-1} = \tilde{V}^{-1}\tilde{U} .$$

The first result is analogous to Theorem 2.1; the proof is omitted.

Lemma 1. The following are equivalent statements about K:

(i) K stabilizes G,

(ii) $\begin{bmatrix} M & U \\ N & V \end{bmatrix}^{-1} \in \mathbf{RH}_\infty$,

(iii) $\begin{bmatrix} \tilde{V} & -\tilde{U} \\ -\tilde{N} & \tilde{M} \end{bmatrix}^{-1} \in \mathbf{RH}_\infty$.

The main result of this chapter is the following.

Theorem 1. The set of all (proper real-rational) K's stabilizing G is parametrized by the formulas

$$K = (Y - MQ)(X - NQ)^{-1} \tag{2}$$

$$= (\tilde{X} - Q\tilde{N})^{-1}(\tilde{Y} - Q\tilde{M}) \tag{3}$$

$$Q \in \mathbf{RH}_\infty .$$

Proof. Let's first prove equality (3). Let $Q \in \mathbf{RH}_\infty$. From (1) we have

$$\begin{bmatrix} I & Q \\ 0 & I \end{bmatrix} \begin{bmatrix} \tilde{X} & -\tilde{Y} \\ -\tilde{N} & \tilde{M} \end{bmatrix} \begin{bmatrix} M & Y \\ N & X \end{bmatrix} \begin{bmatrix} I & -Q \\ 0 & I \end{bmatrix} = I$$

so that

$$\begin{bmatrix} \tilde{X} - Q\tilde{N} & -(\tilde{Y} - Q\tilde{M}) \\ -\tilde{N} & \tilde{M} \end{bmatrix} \begin{bmatrix} M & Y - MQ \\ N & X - NQ \end{bmatrix} = I . \tag{4}$$

Equating the (1,2)-blocks on each side in (4) gives

$$(\tilde{X} - Q\tilde{N})(Y - MQ) = (\tilde{Y} - Q\tilde{M})(X - NQ) ,$$

which is equivalent to (3).

Next, we show that if K is given by (2), it stabilizes G. Define

$$U := Y - MQ \ , \quad V := X - NQ$$

$$\tilde{U} := \tilde{Y} - Q\tilde{M} \ , \quad \tilde{V} := \tilde{X} - Q\tilde{N}$$

o get from (4) that

$$\begin{bmatrix} \tilde{V} & -\tilde{U} \\ -\tilde{N} & \tilde{M} \end{bmatrix} \begin{bmatrix} M & U \\ N & V \end{bmatrix} = I \ . \tag{5}$$

t follows from (5) that U, V are right-coprime and $\tilde{U}$, $\tilde{V}$ are left-coprime Lemma 3.2). Also from (5)

$$\begin{bmatrix} M & U \\ N & V \end{bmatrix}^{-1} \in \mathbf{RH}_\infty \ .$$

So from Lemma 1 K stabilizes G.

Finally, suppose K stabilizes G. We must show K satisfies (2) for some Q ı $\mathbf{RH}_\infty$. Let $K = UV^{-1}$ be a right-coprime factorization. From (1) and defining) $:= \tilde{M}V - \tilde{N}U$ we have

$$\begin{bmatrix} \tilde{X} & -\tilde{Y} \\ -\tilde{N} & \tilde{M} \end{bmatrix} \begin{bmatrix} M & U \\ N & V \end{bmatrix} = \begin{bmatrix} I & \tilde{X}U - \tilde{Y}V \\ 0 & D \end{bmatrix} . \tag{6}$$

he two matrices on the left in (6) have inverses in $\mathbf{RH}_\infty$, the second by Lemma . Hence $D^{-1} \in \mathbf{RH}_\infty$. Define

$$Q := -(\tilde{X}U - \tilde{Y}V)D^{-1} \ ,$$

that (6) becomes

$$\begin{bmatrix} \tilde{X} & -\tilde{Y} \\ -\tilde{N} & \tilde{M} \end{bmatrix} \begin{bmatrix} M & U \\ N & V \end{bmatrix} = \begin{bmatrix} I & -QD \\ 0 & D \end{bmatrix} . \tag{7}$$

re-multiply (7) by

$$\begin{bmatrix} M & Y \\ N & X \end{bmatrix}$$

ıd use (1) to get

$$\begin{bmatrix} M & U \\ N & V \end{bmatrix} = \begin{bmatrix} M & Y \\ N & X \end{bmatrix} \begin{bmatrix} I & -QD \\ 0 & D \end{bmatrix} .$$

Therefore

$$\begin{bmatrix} U \\ V \end{bmatrix} = \begin{bmatrix} (Y - MQ)D \\ (X - NQ)D \end{bmatrix} .$$

Substitute this into $K = UV^{-1}$ to get (2). □

As a special case suppose G is already stable, i.e. $G \in \mathbf{RH}_\infty$. Then in (1) we may take

$$N = \tilde{N} = G$$

$$\tilde{X} = M = I \, , \quad X = \tilde{M} = I$$

$$Y = 0 \, , \quad \tilde{Y} = 0 \, ,$$

in which case the formulas in the theorem become simply

$$K = -Q(I - GQ)^{-1}$$

$$= -(I - QG)^{-1}Q \, .$$

There is an interpretation of Q in this case: $-Q$ equals the transfer matrix from v_2 to u in Figure 1 (check this).

We can now explain the idea behind the choice (1.7) of X, Y, $\tilde{X}$, $\tilde{Y}$ in Section 1. Recall that the state-space equations for G were

$$\dot{x} = Ax + Bu$$

$$y = Cx + Du \, ,$$

that

$$A_F := A + BF \, , \quad A_H := A + HC$$

were stable, and that we defined

$$B_H := B + HD \, , \quad C_F := C + DF \, .$$

Let's find a stabilizing K by observer theory. The familiar state-space equations for K are

$$\dot{\hat{x}} = A\hat{x} + Bu + H(C\hat{x} + Du - y)$$

$$u = F\hat{x} \, ,$$

or equivalently

$$\dot{\hat{x}} = \hat{A}\hat{x} + \hat{B}y$$

$$u = \hat{C}\hat{x} ,$$

where

$$\hat{A} := A + BF + HC + HDF = A_F + HC_F$$

$$\hat{B} := -H$$

$$\hat{C} := F .$$

Thus in terms of our data structure

$$K(s) = [\hat{A}, \hat{B}, \hat{C}, 0] .$$

By observer theory K stabilizes G.

Now find coprime factorizations of K in the same way as we found coprime factorizations of G in Section 1. To get a right-coprime factorization $K = YX^{-1}$ we first choose $\hat{F}$ so that $\hat{A}_F := \hat{A} + \hat{B}\hat{F}$ is stable. It is convenient to take $\hat{F} := C_F$, so that $\hat{A}_F = A_F$. By analogy with (1.6) we get $K = YX^{-1}$, where

$$X(s) := [\hat{A}_F, \hat{B}, \hat{F}, I]$$

$$= [A_F, -H, C_F, I]$$

$$Y(s) := [\hat{A}_F, \hat{B}, \hat{C}, 0]$$

$$= [A_F, -H, F, 0] .$$

A similar derivation leads to a left-coprime factorization $K = \tilde{X}^{-1}\tilde{Y}$, where

$$\tilde{X}(s) := [A_H, -B_H, F, I]$$

$$\tilde{Y}(s) := [A_H, -H, F, 0] .$$

These formulas coincide with (1.7).

By Lemma 1 we know that

$$\begin{bmatrix} M & Y \\ N & X \end{bmatrix}^{-1} \in \mathbf{RH}_\infty$$

and

$$\begin{bmatrix} \tilde{X} & -\tilde{Y} \\ -\tilde{N} & \tilde{M} \end{bmatrix}^{-1} \in \mathbf{RH}_\infty .$$

Hence the product

$$\begin{bmatrix} \tilde{X} & -\tilde{Y} \\ -\tilde{N} & \tilde{M} \end{bmatrix} \begin{bmatrix} M & Y \\ N & X \end{bmatrix}$$

must be invertible in $\mathbf{RH}_\infty$. The only surprise is that the product equals the identity matrix, as is verified by algebraic manipulation.

Exercise 1. In Figure 4 suppose $G(s) = \frac{1}{s(s-1)}$. Consider a controller of the form

$$K = \frac{-Q}{1-GQ}$$

where Q is real-rational. Find necessary and sufficient conditions on Q in order that K stabilize G.

4.5 Closed-Loop Transfer Matrices

Now we return to the standard set-up of Figure 1, Chapter 3. Theorem 4.1 gives every stabilizing K as a transformation of a free parameter Q in $\mathbf{RH}_\infty$. The objective in this section is to find the transfer matrix from w to z in terms of Q.

In the previous section we dropped the subscripts on G_{22}; now we must restore them. Bring in a doubly-coprime factorization of G_{22}:

$$G_{22} = N_2 M_2^{-1} = \tilde{M}_2^{-1} \tilde{N}_2$$

$$\begin{bmatrix} \tilde{X}_2 & -\tilde{Y}_2 \\ -\tilde{N}_2 & \tilde{M}_2 \end{bmatrix} \begin{bmatrix} M_2 & Y_2 \\ N_2 & X_2 \end{bmatrix} = I . \tag{1}$$

Then the formula for K is

$$K = (Y_2 - M_2 Q)(X_2 - N_2 Q)^{-1} \tag{2a}$$

$$= (\tilde{X}_2 - Q\tilde{N}_2)^{-1}(\tilde{Y}_2 - Q\tilde{M}_2) . \tag{2b}$$

Now define

$$T_1 := G_{11} + G_{12}M_2\tilde{Y}_2G_{21} \tag{3a}$$

$$T_2 := G_{12}M_2 \tag{3b}$$

$$T_3 := \tilde{M}_2G_{21} . \tag{3c}$$

Theorem 1. The matrices T_i $(i = 1\text{–}3)$ belong to $\mathbf{RH}_\infty$. With K given by (2) the transfer matrix from w to z equals $T_1 - T_2QT_3$.

Proof. The first statement follows from the realizations to be given below. For the second statement we have

$$z = [G_{11} + G_{12}(I - KG_{22})^{-1}KG_{21}]w . \tag{4}$$

Substitute $G_{22} = N_2M_2^{-1}$ and (2b) into $(I - KG_{22})^{-1}$ and use (1) to get

$$(I - KG_{22})^{-1} = M_2(\tilde{X}_2 - Q\tilde{N}_2) .$$

Thus from (2b) again

$$(I - KG_{22})^{-1}K = M_2(\tilde{Y}_2 - Q\tilde{M}_2) .$$

Substitute this into (4) and use the definitions of T_i to get

$$z = (T_1 - T_2QT_3)w . \quad \square$$

For computations it is useful to have explicit realizations of the transfer matrices T_i $(i = 1\text{–}3)$. Start with a minimal realization of G :

$$G(s) = [A, B, C, D] .$$

Since the input and output of G are partitioned as

$$\begin{bmatrix} w \\ u \end{bmatrix}, \quad \begin{bmatrix} z \\ y \end{bmatrix},$$

the matrices B, C, and D have corresponding partitions:

$$B = [B_1 \quad B_2]$$

$$C = \begin{bmatrix} C_1 \\ C_2 \end{bmatrix}$$

$$D = \begin{bmatrix} D_{11} & D_{12} \\ D_{21} & D_{22} \end{bmatrix}.$$

Then

$$G_{ij}(s) = [A, B_j, C_i, D_{ij}], \quad i,j=1,2.$$

Note that $D_{22}=0$ because G_{22} is strictly proper. It can be proved that stabilizability of G (an assumption from Section 3) implies that (A, B_2) is stabilizable and (C_2, A) is detectable.

Next, find a doubly-coprime factorization of G_{22} as developed in Section 1. For this choose F and H so that

$$A_F := A + B_2F, \quad A_H := A + HC_2$$

are stable. Then the formulas are as follows:

$$M_2(s) = [A_F, B_2, F, I]$$

$$N_2(s) = [A_F, B_2, C_2, 0]$$

$$\tilde{M}_2(s) = [A_H, H, C_2, I]$$

$$\tilde{N}_2(s) = [A_H, B_2, C_2, 0]$$

$$X_2(s) = [A_F, -H, C_2, I]$$

$$Y_2(s) = [A_F, -H, F, 0]$$

$$\tilde{X}_2(s) = [A_H, -B_2, F, I]$$

$$\tilde{Y}_2(s) = [A_H, -H, F, 0].$$

Finally, substitution into (3) yields the following realizations:

$$T_1(s) = [\underline{A}, \underline{B}, \underline{C}, D_{11}]$$

$$\underline{A} = \begin{bmatrix} A_F & -B_2F \\ 0 & A_H \end{bmatrix}$$

$$\underline{B} = \begin{bmatrix} B_1 \\ B_1 + HD_{21} \end{bmatrix}$$

$$\underline{C} = [C_1 + D_{12}F \quad -D_{12}F]$$

$$T_2(s) = [A_F, B_2, C_1 + D_{12}F, D_{12}]$$

$$T_3(s) = [A_H, B_1 + HD_{21}, C_2, D_{21}] \; .$$

It can be observed that $T_i \in \mathbf{RH}_\infty$ ($i = 1$–3), as claimed in Theorem 1. For example, this is how the realization of T_2 is obtained:

$$T_2(s) = G_{12}(s)M_2(s)$$

$$= [A, B_2, C_1, D_{12}] \times [A_F, B_2, F, I]$$

$$= \left[\begin{bmatrix} A & B_2F \\ 0 & A_F \end{bmatrix}, \begin{bmatrix} B_2 \\ B_2 \end{bmatrix}, [C_1 \quad D_{12}F], D_{12} \right]$$

$$= \left[\begin{bmatrix} A & 0 \\ 0 & A_F \end{bmatrix}, \begin{bmatrix} 0 \\ B_2 \end{bmatrix}, [C_1 \quad C_1 + D_{12}F], D_{12} \right] \tag{5}$$

$$= [A_F, B_2, C_1 + D_{12}F, D_{12}] \; .$$

Similarity transformation by $\begin{bmatrix} I & I \\ 0 & I \end{bmatrix}$ was used in (5).

Example 1.

Consider the tracking example of Chapter 3 with

$$P(s) = \frac{s-1}{s(s-2)}$$

$$W(s) = \frac{s+1}{10s+1}$$

and $\rho = 1$. We have

$$G = \begin{bmatrix} G_{11} & G_{12} \\ G_{21} & G_{22} \end{bmatrix}$$

$$G_{11}(s) = \begin{bmatrix} \dfrac{s+1}{10s+1} \\ 0 \end{bmatrix}, \quad G_{12}(s) = \begin{bmatrix} -\dfrac{s-1}{s(s-2)} \\ 1 \end{bmatrix}$$

$$G_{21}(s) = G_{11}(s), \quad G_{22}(s) = \begin{bmatrix} 0 \\ \dfrac{s-1}{s(s-2)} \end{bmatrix} .$$

A minimal realization of G is

$$G(s) = [A, B, C, D]$$

$$A = \begin{bmatrix} -.1 & 0 & 0 \\ 0 & 2 & 0 \\ 0 & 1 & 0 \end{bmatrix}, \quad B = \begin{bmatrix} 1 & 0 \\ 0 & 1 \\ 0 & 0 \end{bmatrix}$$

$$C = \begin{bmatrix} .09 & -1 & 1 \\ 0 & 0 & 0 \\ .09 & 0 & 0 \\ 0 & 1 & -1 \end{bmatrix}, \quad D = \begin{bmatrix} .1 & 0 \\ 0 & 1 \\ .1 & 0 \\ 0 & 0 \end{bmatrix}.$$

For F and H we may take

$$F = [0 \ \ -3 \ \ -1]$$

$$H = \begin{bmatrix} 0 & 0 \\ 0 & -9 \\ 0 & -5 \end{bmatrix}.$$

Then

$$A_F = \begin{bmatrix} -.1 & 0 & 0 \\ 0 & -1 & -1 \\ 0 & 1 & 0 \end{bmatrix}, \quad A_H = \begin{bmatrix} -.1 & 0 & 0 \\ 0 & -7 & 9 \\ 0 & -4 & 5 \end{bmatrix}$$

$$\underline{A} = \begin{bmatrix} -.1 & 0 & 0 & 0 & 0 & 0 \\ 0 & -1 & -1 & 0 & 3 & 1 \\ 0 & 1 & 0 & 0 & 0 & 0 \\ 0 & 0 & 0 & -.1 & 0 & 0 \\ 0 & 0 & 0 & 0 & -7 & 9 \\ 0 & 0 & 0 & 0 & -4 & 5 \end{bmatrix}, \quad \underline{B} = \begin{bmatrix} 1 \\ 0 \\ 0 \\ 1 \\ 0 \\ 0 \end{bmatrix}$$

$$\underline{C} = \begin{bmatrix} .09 & -1 & 1 & 0 & 0 & 0 \\ 0 & -3 & -1 & 0 & 3 & 1 \end{bmatrix}.$$

Finally,

$$T_1(s) = \begin{bmatrix} \dfrac{s+1}{10s+1} \\ 0 \end{bmatrix}$$

$$T_2(s) = \begin{bmatrix} -\dfrac{s-1}{s^2+s+1} \\ \dfrac{s(s-2)}{s^2+s+1} \end{bmatrix}$$

$$T_3(s) = T_1(s) .$$

Exercise 1. It's desired to find a K which stabilizes G and makes the dc gain from w to z equal to zero (asymptotic rejection of steps). Give an example of a stabilizable G for which no such K exists. Find necessary and sufficient conditions (in terms of T_i, $i=1$–3) for such K to exist.

The results of this chapter can be summarized as follows. The matrix G is assumed to be proper, with G_{22} strictly proper. Also, G is assumed to be stabilizable. The formula (2) parametrizes all K's which stabilize G. In terms of the parameter Q the transfer matrix from w to z equals $T_1-T_2QT_3$. Such a function of Q is called *affine*.

In view of these results the standard problem can be solved as follows: First, find a Q in $\mathbf{RH}_\infty$ to minimize $||T_1-T_2QT_3||_\infty$, i.e. solve the model-matching problem of Chapter 3. Then obtain a controller K by substituting Q into (2).

Notes and References

The material of this chapter is based on Doyle (1984). Earlier relevant references are Chang and Pearson (1978) and Pernebo (1981); a more general treatment is given in Nett (1985).

As a general reference for the material of this chapter see Vidyasagar (1985a). The idea of doing coprime factorization over $\mathbf{RH}_\infty$ is due to Vidyasagar (1972), but the idea was first fully exploited by Desoer *et al.* (1980). State-space formulas for coprime factorizations were first developed by Khargonekar and Sontag (1982). The state-space formulas in Section 1 are from Nett *et al.* (1984). The important parametrization of Theorem 4.1 is due to Youla *et al.* (1976) as modified by Desoer *et al.* (1980). Finally, see Minto (1985) for a comprehensive treatment of stability theory by state-space methods.

CHAPTER 5

BACKGROUND MATHEMATICS: OPERATORS

The purposes of this chapter are to introduce some basic definitions about operators on Hilbert space and to study in some detail a certain type of operator, namely, a Hankel operator.

5.1 Hankel Operators

Let $\mathbf{X}$ and $\mathbf{Y}$ be Hilbert spaces and let Φ be a linear function from $\mathbf{X}$ to $\mathbf{Y}$. The function Φ is *bounded* if there exists a real number a such that

$$||\Phi x|| \leq a\, ||x||\,, \quad x \in \mathbf{X}\,.$$

The least such a is called the *norm* of Φ, denoted $||\Phi||$. The following equations are not hard to derive:

$$||\Phi|| = \sup\,\{||\Phi x|| : ||x|| \leq 1\}$$

$$= \sup\,\{||\Phi x|| : ||x|| = 1\}\,.$$

Such a bounded linear function is called an *operator.*

Example 1.

The Fourier transform, denoted Υ, is an operator from $\mathbf{L}_2(-\infty,\infty)$ to $\mathbf{L}_2$. Theorem 2.4.1 says that its norm equals 1.

Example 2.

Recall from Section 2.3 the direct sum

$$\mathbf{L}_2(-\infty,\infty) = \mathbf{L}_2(-\infty,0] \oplus \mathbf{L}_2[0,\infty)\,.$$

Each function f in $\mathbf{L}_2(-\infty,\infty)$ has a unique decomposition $f = f_1 + f_2$ with $f_1 \in \mathbf{L}_2(-\infty,0]$ and $f_2 \in \mathbf{L}_2[0,\infty)$:

$$f_1(t) = f(t), \quad f_2(t) = 0, \qquad t \leq 0$$

$$f_1(t) = 0, \quad f_2(t) = f(t), \qquad t > 0 .$$

The function $f \rightarrow f_1$ from $\mathbf{L}_2(-\infty,\infty)$ to $\mathbf{L}_2(-\infty,0]$ is an operator, the *orthogonal projection* of $\mathbf{L}_2(-\infty,\infty)$ onto $\mathbf{L}_2(-\infty,0]$. It's easy to prove that its norm equals 1.

In the same way we have

$$\mathbf{L}_2 = \mathbf{H}_2^{\perp} \oplus \mathbf{H}_2 .$$

The orthogonal projection from $\mathbf{L}_2$ onto $\mathbf{H}_2^{\perp}$ will be denoted Π_1 and from $\mathbf{L}_2$ onto $\mathbf{H}_2$ by Π_2.

Example 3.

Let $F \in \mathbf{L}_{\infty}$ and define the function Λ_F from $\mathbf{L}_2$ to $\mathbf{L}_2$ via

$$\Lambda_F \, g := Fg .$$

Thus the action of Λ_F is multiplication by F. Obviously Λ_F is linear. Theorem 2.4.2 says that $||\Lambda_F|| = ||F||_{\infty}$, so Λ_F is bounded. This operator is called a *Laurent operator* and F is called its *symbol*; so Λ_F is the Laurent operator with symbol F.

A related operator is $\Lambda_F \mid \mathbf{H}_2$, the *restriction* of Λ_F to $\mathbf{H}_2$, which maps $\mathbf{H}_2$ to $\mathbf{L}_2$. Theorem 2.4.2 says that its norm also equals $||F||_{\infty}$.

Observe that if $F \in \mathbf{H}_{\infty}$ then, also by Theorem 2.4.2,

$$\Lambda_F \, \mathbf{H}_2 \subset \mathbf{H}_2 .$$

The converse is true too: if $\Lambda_F \, \mathbf{H}_2 \subset \mathbf{H}_2$, then $F \in \mathbf{H}_{\infty}$. (The rational version of this result will be proved in Lemma 8.3.1.)

Example 4.

This is the time-domain analog of the previous example. Recall that convolution in the time-domain corresponds to multiplication in the frequency-domain. Suppose $F(s)$ is a matrix-valued function which is analytic in a vertical strip containing the imaginary axis and which belongs to $\mathbf{L}_{\infty}$. Taking the region of

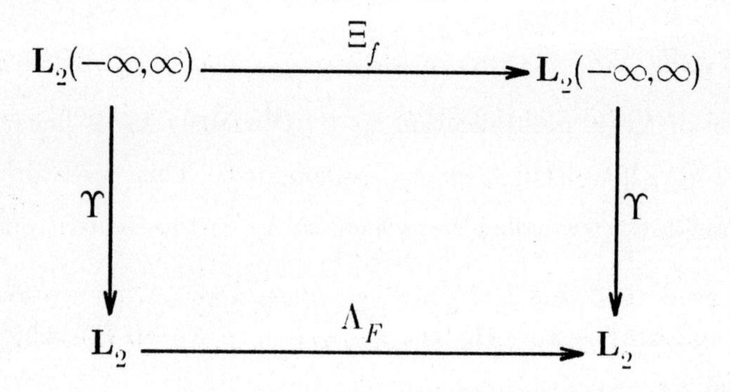

Figure 5.1.1. Example 4

convergence to be this strip, let $f(t)$ denote the inverse bilateral Laplace transform of $F(s)$. Now define the convolution operator Ξ_f from $\mathbf{L}_2(-\infty,\infty)$ to $\mathbf{L}_2(-\infty,\infty)$ via

$$y = \Xi_f u$$

$$y(t) = \int_{-\infty}^{\infty} f(t-\tau)u(\tau)d\tau .$$

This system is linear, but not necessarily causal because $f(t)$ may not equal zero for negative time. Note that the system is causal iff Ξ_f maps $\mathbf{L}_2[0,\infty)$ into $\mathbf{L}_2[0,\infty)$, i.e. "Ξ_f leaves the future invariant". The operators Ξ_f and Λ_F are intimately related via the Fourier transform. The relationship is exhibited in the commutative diagram of Figure 1.

Example 5.

Again let $F \in \mathbf{L}_\infty$. The *Toeplitz operator with symbol* F, denoted Θ_F, maps $\mathbf{H}_2$ to $\mathbf{H}_2$ and is defined as follows: for each g in $\mathbf{H}_2$, $\Theta_F g$ equals the orthogonal projection of Fg onto $\mathbf{H}_2$. Thus

$$\Theta_F = \Pi_2 \Lambda_F \mid \mathbf{H}_2 .$$

The relevant commutative diagram is Figure 2. As a concrete example consider the scalar-valued function $F(s) = 1/(s-1)$ in $\mathbf{RL}_\infty$. For g in $\mathbf{H}_2$ we have

$$Fg = g_1 + g_2$$

$$g_1 \in \mathbf{H}_2^\perp, \quad g_2 \in \mathbf{H}_2$$

$$g_1(s) = g(1)/(s-1)$$

$$g_2(s) = [g(s)-g(1)]/(s-1) .$$

Thus Θ_F maps g to g_2.

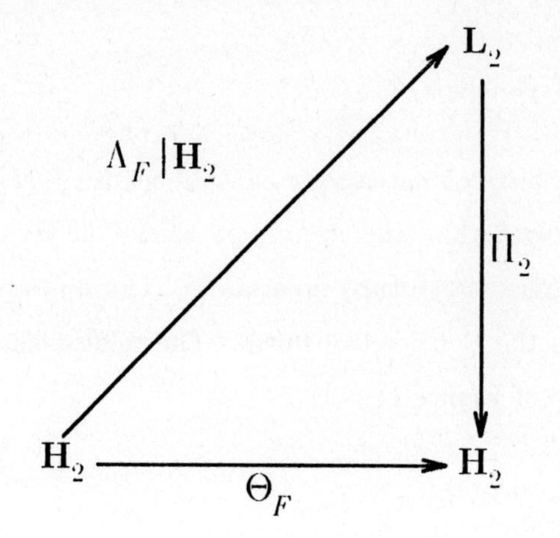

Figure 5.1.2. Example 5

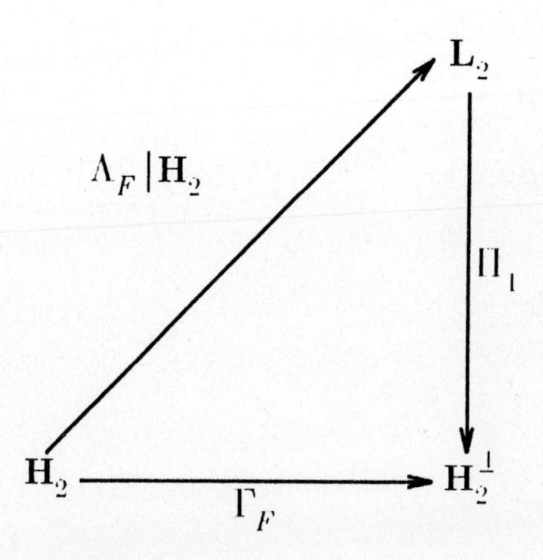

Figure 5.1.3. Example 6

Example 6.

For F in $\mathbf{L}_\infty$ the *Hankel operator with symbol F*, denoted Γ_F, maps $\mathbf{H}_2$ to $\mathbf{H}_2^\perp$ and is defined as

$$\Gamma_F := \Pi_1 \Lambda_F \mid \mathbf{H}_2 .$$

The corresponding commutative diagram is Figure 3. For the example $F(s) = 1/(s-1)$, Γ_F maps $g(s)$ in $\mathbf{H}_2$ to $g(1)/(s-1)$ in $\mathbf{H}_2^\perp$. Note that $\Gamma_F = 0$ if $F \in \mathbf{H}_\infty$.

The relationship between the three operators Λ_F, Θ_F, and Γ_F can be described as follows. We have

$$\Lambda_F : \mathbf{H}_2^\perp \oplus \mathbf{H}_2 \rightarrow \mathbf{H}_2^\perp \oplus \mathbf{H}_2 ,$$

and correspondingly we can regard Λ_F as a 2×2 matrix with operator entries:

$$\Lambda_F = \begin{bmatrix} \Lambda_{11} & \Lambda_{12} \\ \Lambda_{21} & \Lambda_{22} \end{bmatrix} .$$

For example

$$\Lambda_{11} = \Pi_1 \Lambda_F \mid \mathbf{H}_2^\perp .$$

It follows from the definitions that $\Lambda_{12} = \Gamma_F$ and $\Lambda_{22} = \Theta_F$. Thus

$$\Lambda_F \mid \mathbf{H}_2 = \begin{bmatrix} \Gamma_F \\ \Theta_F \end{bmatrix} .$$

Example 7.

In this example we study the Hankel operator with the special symbol

$$F(s) = [A, B, C, 0] ,$$

where A is antistable (all eigenvalues in Re $s > 0$). Suppose A is $n \times n$. Such F belongs to $\mathbf{RL}_\infty$. The inverse bilateral Laplace transform of $F(s)$ is

$$f(t) = -Ce^{At}B , \quad t < 0$$

$$f(t) = 0 , \quad t \geq 0 .$$

The time-domain analog of the Hankel operator, denoted Γ_f, maps a function u in $\mathbf{L}_2[0,\infty)$ to the function y in $\mathbf{L}_2(-\infty,0]$ defined by

$$y(t) = \int_0^\infty f(t-\tau)u(\tau)d\tau, \quad t<0$$

$$= -Ce^{At} \int_0^\infty e^{-A\tau}Bu(\tau)d\tau, \quad t<0 . \tag{1}$$

Define two auxiliary operators: the *controllability operator*

$$\Psi_c : \mathbf{L}_2[0,\infty) \rightarrow \mathbf{C}^n$$

$$\Psi_c u := -\int_0^\infty e^{-A\tau}Bu(\tau)d\tau$$

and the *observability operator*

$$\Psi_o : \mathbf{C}^n \rightarrow \mathbf{L}_2(-\infty,0]$$

$$(\Psi_o x)(t) := Ce^{At}x, \quad t<0 .$$

From (1) we have that

$$\Gamma_f = \Psi_o \Psi_c .$$

Exercise 1. Show that Ψ_c is surjective if (A,B) is controllable and that Ψ_o is injective if (C,A) is observable.

There is a systemic interpretation of Γ_f in terms of the usual state-space equations

$$\dot{x} = Ax + Bu \tag{2}$$

$$y = Cx . \tag{3}$$

To see the action of Γ_f, solve these equations in the following way. First, apply an input u in $\mathbf{L}_2[0,\infty)$ to equation (2) with initial condition $x(0)=x_0$ and such that $x(t)$ is bounded on $[0,\infty)$. Then

$$x(t) = e^{At}x_0 + e^{At}\int_0^t e^{-A\tau}Bu(\tau)d\tau, \quad t\geq 0$$

so that

$$x_0 = -\int_0^\infty e^{-A\tau}Bu(\tau)d\tau$$

$$= \Psi_c u \ .$$

Now solve (2) and (3) backwards in time starting at $t = 0$ and noting that $u(t) = 0$ for $t < 0$. The solution is

$$y(t) = Ce^{At} x_0$$

$$= (\Psi_o x_0)(t), \quad t < 0 \ .$$

In this way Γ_f maps future input to initial state to past output.

We shall need the concept of the *adjoint* of an operator Φ from $\mathbf{X}$ to $\mathbf{Y}$, two Hilbert spaces: it's the unique operator Φ^* from $\mathbf{Y}$ to $\mathbf{X}$ satisfying

$$<\Phi x, y> = <x, \Phi^* y>, \quad x \in \mathbf{X}, \quad y \in \mathbf{Y} \ .$$

The operator $\Phi^* \Phi$ from $\mathbf{X}$ to $\mathbf{X}$ is self-adjoint, i.e. it equals its adjoint. The norms of Φ and $\Phi^* \Phi$ are related as follows:

$$||\Phi||^2 = ||\Phi^* \Phi|| \ . \tag{4}$$

The adjoint of a Laurent operator Λ_F can be obtained explicitly as follows. Introduce the notation

$$F^{\sim}(j\omega) := F(j\omega)^* \ , \tag{5}$$

where * here denotes complex-conjugate transpose. If $F \in \mathbf{RL}_\infty$, then we shall interpret $F^{\sim}$ as

$$F^{\sim}(s) := F(-s)^T \ ,$$

which is consistent with (5). If g and h belong to $\mathbf{L}_2$, then

$$<\Lambda_F g, h> = <g, \Lambda_F^* h>$$

and

$$<\Lambda_F g, h> = (2\pi)^{-1} \int_{-\infty}^{\infty} g(j\omega)^* F(j\omega)^* h(j\omega) d\omega$$

$$= <g, F^{\sim} h> \ .$$

We conclude that Λ_F^* equals the Laurent operator with symbol $F^{\sim}$.

Similarly, the adjoint of a Hankel operator Γ_F can be characterized as follows. Let $g \in \mathbf{H}_2$ and $h \in \mathbf{H}_2^\perp$. Then

$$\begin{aligned} < g, \Gamma_F^* h > &= < \Gamma_F g, h > \\ &= < \Pi_1 F g, h > \\ &= < F g, h > \quad \text{because } \Pi_2 F g \perp h \\ &= < g, F^\sim h > \\ &= < g, \Pi_2 F^\sim h > \quad \text{because } g \perp \Pi_1 F^\sim h \ . \end{aligned}$$

We conclude that

$$\Gamma_F^* = \Pi_2 \Lambda_F^* \mid \mathbf{H}_2^\perp .$$

For example if $F(s) = (s-1)^{-1}$, then Γ_F^* maps $h(s)$ in $\mathbf{H}_2^\perp$ to $-h(-1)/(s+1)$ in $\mathbf{H}_2$. (Verify.)

Exercise 2. Show that the adjoints of Ψ_c and Ψ_o are as follows:

$$\Psi_c^* : \mathbf{C}^n \rightarrow \mathbf{L}_2[0,\infty)$$

$$(\Psi_c^* x)(t) = -B^T e^{-A^T t} x \ , \quad t \geq 0;$$

$$\Psi_o^* : \mathbf{L}_2(-\infty,0] \rightarrow \mathbf{C}^n$$

$$\Psi_o^* y = \int_{-\infty}^{0} e^{A^T t} C^T y(t) dt \ .$$

The *rank* of an operator $\Phi : \mathbf{X} \rightarrow \mathbf{Y}$ is the dimension of the closure of its image space $\Phi\mathbf{X}$.

Our interest is in Hankel operators with real-rational symbols, Example 7 being a special case.

Theorem 1. If $F \in \mathbf{RL}_\infty$, then Γ_F has finite rank.

Proof. There is a unique factorization (by partial-fraction expansion, for example) $F = F_1 + F_2$, where F_1 is strictly proper and analytic in Re $s \leq 0$ and F_2 is proper and analytic in Re $s \geq 0$, i.e. $F_2 \in \mathbf{RH}_\infty$. Since $\Gamma_F = \Gamma_{F_1}$, we might as well assume at the start that F is strictly proper and analytic in Re $s \leq 0$. Introduce a minimal realization:

$$F(s) = [A, B, C, 0] .$$

The operator Γ_F and its time-domain analog have equal ranks. As in Example 7 the latter operator equals $\Psi_o \Psi_c$. By controllability and observability Ψ_c is surjective and Ψ_o is injective. Hence $\Psi_o \Psi_c$ has rank n, so Γ_F does too. □

We need another definition. Let Φ be an operator from **X** to **X**, a Hilbert space. A complex number λ is an *eigenvalue* of Φ if there is a nonzero x in **X** satisfying

$$\Phi x = \lambda x .$$

Then x is an *eigenvector corresponding to* λ. In general an operator may not have any eigenvalues!

For the remainder of this section let $F \in \mathbf{RL}_\infty$. The self-adjoint operator $\Gamma_F^* \Gamma_F$ maps $\mathbf{H}_2$ to itself and its rank is finite by Theorem 1. This property guarantees that it does in fact have eigenvalues. We state without proof the following fact.

Theorem 2. The eigenvalues of $\Gamma_F^* \Gamma_F$ are real and nonnegative and the largest of them equals $\|\Gamma_F^* \Gamma_F\|$.

This theorem together with (4) says that $\|\Gamma_F\|$ equals the square root of the largest eigenvalue of $\Gamma_F^* \Gamma_F$. So we could compute $\|\Gamma_F\|$ if we could compute the eigenvalues of $\Gamma_F^* \Gamma_F$. How to do this latter computation is the last topic of this section.

We continue with the notation introduced in Example 7 and the proof of Theorem 1. The self-adjoint operators $\Psi_c \Psi_c^*$ and $\Psi_o^* \Psi_o$ map $\mathbf{C}^n$ to itself. Thus they have matrix representations with respect to the standard basis on $\mathbf{C}^n$. Define the *controllability* and *observability gramians*

$$L_c := \int_0^\infty e^{-At} BB^T e^{-A^T t} \, dt \tag{6}$$

$$L_o := \int_0^\infty e^{-A^T t} C^T C e^{-At} \, dt \ . \tag{7}$$

It is routine to show that L_c and L_o are the unique solutions of the Lyapunov equations

$$AL_c + L_c A^T = BB^T \tag{8}$$

$$A^T L_o + L_o A = C^T C \ . \tag{9}$$

Exercise 3. Prove that the matrix representations of $\Psi_c \Psi_c^*$ and $\Psi_o^* \Psi_o$ are L_c and L_o respectively.

Theorem 3. The operator $\Gamma_F^* \Gamma_F$ and the matrix $L_c L_o$ share the same nonzero eigenvalues.

Proof. Let λ be a nonzero eigenvalue of $\Gamma_F^* \Gamma_F$. It's easy to show that λ is also an eigenvalue of the time-domain analog of $\Gamma_F^* \Gamma_F$, which equals $\Psi_c^* \Psi_o^* \Psi_o \Psi_c$. Hence there exists a nonzero u in $\mathbf{L}_2[0,\infty)$ satisfying

$$\Psi_c^* \Psi_o^* \Psi_o \Psi_c u = \lambda u \ . \tag{10}$$

Pre-multiply (10) by Ψ_c and define $x := \Psi_c u$ to get

$$L_c L_o x = \lambda x \ . \tag{11}$$

If x, i.e. $\Psi_c u$, were to equal zero, then so would λu from (10). This is not possible (both λ and u are nonzero), so x is an eigenvector of $L_c L_o$ and λ is an eigenvalue.

Conversely, let λ be a nonzero eigenvalue of $L_c L_o$ and x a corresponding eigenvector. Pre-multiply (11) by $\Psi_c^* L_o$ and define $u := \Psi_c^* L_o x$ to get (10). The function u is nonzero because x is nonzero and Ψ_c^* and L_o are injective. Therefore λ is an eigenvalue of $\Psi_c^* \Psi_o^* \Psi_o \Psi_c$, hence of $\Gamma_F^* \Gamma_F$. □

In summary, the norm of Γ_F for F in $\mathbf{RL}_\infty$ can be computed as follows. First, find a minimal realization (A, B, C) of the antistable part of $F(s)$, i.e.

$$F(s) = [A, B, C, 0] + (\text{a matrix in } \mathbf{RH}_\infty) .$$

Next, solve the Lyapunov equations (8) and (9) for L_c and L_o. Then $||\Gamma_F||$ equals the square root of the largest eigenvalue of $L_c L_o$.

5.2 Nehari's Theorem

In this section we look at the problem of finding the distance from an $\mathbf{L}_\infty$-matrix R to $\mathbf{H}_\infty$:

$$\text{dist}(R, \mathbf{H}_\infty) := \inf \{ ||R - X||_\infty : X \in \mathbf{H}_\infty \} .$$

In systemic terms we want to approximate, in $\mathbf{L}_\infty$-norm, a given unstable transfer matrix by a stable one. Nehari's theorem is an elegant solution to this problem.

A lower bound for the distance is easily obtained. Fix X in $\mathbf{H}_\infty$. Then

$$\begin{aligned} ||R - X||_\infty &= ||\Lambda_R - \Lambda_X|| \\ &\geq ||\Pi_1(\Lambda_R - \Lambda_X) \mid \mathbf{H}_2|| \\ &= ||\Gamma_R - \Gamma_X|| \\ &= ||\Gamma_R|| . \end{aligned}$$

The last equality is due to the fact that $\Gamma_X = 0$. Thus $||\Gamma_R||$ is a lower bound for the distance from R to $\mathbf{H}_\infty$. In fact it equals the distance.

Theorem 1. There exists a closest $\mathbf{H}_\infty$-matrix X to a given $\mathbf{L}_\infty$-matrix R, and $||R - X|| = ||\Gamma_R||$.

In general there are many X's nearest R. Interpreted in the time-domain Theorem 1 states that the distance from a given noncausal system to the nearest causal one (the systems being linear and time-invariant) equals the norm of the Hankel operator; in other words the norm of the Hankel operator is a measure of noncausality. How to find nearest $\mathbf{H}_\infty$-matrices is the subject of Section 8.3.

Example 1.

As an illustration, let's find the distance from

$$R(s) = \begin{bmatrix} \frac{1}{s^2-1} & 4 \\ \frac{1}{s^2-s+1} & \frac{s+1}{s-1} \end{bmatrix}$$

to $\mathbf{H}_\infty$. First, find the strictly proper antistable part of R. In this simple case partial fraction expansion suffices; a state-space procedure is described in the proof of Theorem 7.3.1. We get

$$R(s) = R_1(s) + R_2(s)$$

$$R_1(s) = \begin{bmatrix} \frac{.5}{s-1} & 0 \\ \frac{1}{s^2-s+1} & \frac{2}{s-1} \end{bmatrix}$$

$$R_2 \in \mathbf{RH}_\infty .$$

Now get a minimal realization:

$$R_1(s) = [A, B, C, 0]$$

$$A = \begin{bmatrix} 2 & -2 & 1 & 0 \\ 1 & 0 & 0 & 0 \\ 0 & 1 & 0 & 0 \\ 0 & 0 & 0 & 1 \end{bmatrix}, \quad B = \begin{bmatrix} 1 & 0 \\ 0 & 0 \\ 0 & 0 \\ 0 & 2 \end{bmatrix}$$

$$C = \begin{bmatrix} .5 & -.5 & .5 & 0 \\ 0 & 1 & -1 & 1 \end{bmatrix}.$$

The solutions of the Lyapunov equations are

$$L_c = \begin{bmatrix} .3333 & 0 & -.1667 & 0 \\ 0 & .1667 & 0 & 0 \\ -.1667 & 0 & .3333 & 0 \\ 0 & 0 & 0 & 2 \end{bmatrix}$$

$$L_o = \begin{bmatrix} .6250 & -1.125 & .6250 & -.3333 \\ -1.125 & 2.625 & -1.625 & 1 \\ .6250 & -1.625 & 1.125 & -.6667 \\ -.3333 & 1 & -.6667 & .5 \end{bmatrix}.$$

Finally, the distance from R to $\mathbf{H}_\infty$ equals the square root of the largest eigenvalue of $L_c L_o$, 1.2695.

Notes and References

An elementary book on operators is Gohberg and Goldberg (1981). Many relevant and interesting results can also be found in Halmos (1982) and Rosenblum and Rovnyak (1985). Theorem 1.1, known as Kronecker's lemma, is a basic fact used in linear system theory. Theorem 1.2 is a special case of a result on compact operators; see for example Theorem 4.4 in Gohberg and Goldberg (1981). Example 1.7 and Theorem 1.3 are from Glover (1984). Theorem 2.1 is a generalization of Nehari's original result (Nehari (1957)); for a (relatively) simple proof see Power (1982).

CHAPTER 6

MODEL-MATCHING THEORY: PART I

The model-matching problem is this: given three matrices T_i in $\mathbf{RH}_\infty$, find a matrix Q in $\mathbf{RH}_\infty$ to minimize $||T_1 - T_2QT_3||_\infty$. This chapter discusses when the problem is solvable and gives a complete solution in the scalar-valued case.

6.1 Existence of a Solution

To each Q in $\mathbf{RH}_\infty$ there corresponds a *model-matching error*, $||T_1 - T_2QT_3||_\infty$. Let α denote the *infimal model-matching error*:

$$\alpha := \inf\{||T_1 - T_2QT_3||_\infty : Q \in \mathbf{RH}_\infty\} . \tag{1}$$

A matrix Q in $\mathbf{RH}_\infty$ satisfying

$$\alpha = ||T_1 - T_2QT_3||_\infty$$

will be called *optimal.*

This section is concerned with the question of when an optimal Q exists. Let's look at a few examples to get a feel for the problem.

Example 1.

The trivial case is when the linear equation

$$T_1 = T_2QT_3 \tag{2}$$

has a solution in $\mathbf{RH}_\infty$. Such a solution is obviously optimal. It is not a difficult problem, but is not germane to this course, to get necessary and sufficient conditions for solvability of (2). Solvability hardly ever occurs in practice, because it means in the standard problem that the exogenous signal w can be completely decoupled from the output signal z .

The remaining examples are of the special case where T_i and Q are scalar-valued. Then there's no need for both T_2 and T_3 since $T_2QT_3=T_2T_3Q$, so we may as well suppose $T_3=1$. In addition, we'll need the following definition: a function F in $\mathbf{RL}_\infty$ is *all-pass* if $|F(j\omega)| =$ constant.

Example 2.

Take T_2 to have one zero in Re $s>0$:

$$T_2(s)=\frac{s-1}{s+1}.$$

Then for every Q in $\mathbf{RH}_\infty$

$$\begin{aligned}||T_1-T_2Q||_\infty &\geq |T_1(1)-T_2(1)Q(1)| \\ &= |T_1(1)|,\end{aligned}$$

so $\alpha\geq|T_1(1)|$. Defining

$$Q:=[T_1-T_1(1)]/T_2,$$

we have $Q\in\mathbf{RH}_\infty$ and

$$||T_1-T_2Q||_\infty=|T_1(1)|,$$

so Q is optimal. Moreover, T_1-T_2Q is all-pass; in fact it's a constant. Thus in this example an optimal Q exists and, furthermore, T_1-T_2Q is all-pass. We'll see later that the optimal Q is unique in this example.

Example 3.

Take

$$T_2(s)=\frac{1}{s+1}.$$

So T_2 has a zero at $s=\infty$ but no (finite) zeros in Re $s\geq0$. For Q in $\mathbf{RH}_\infty$

$$||T_1-T_2Q||_\infty\geq|T_1(\infty)|,$$

so $\alpha\geq|T_1(\infty)|$. As in the previous example, defining

$$Q:=[T_1-T_1(\infty)]/T_2,$$

we have that Q is optimal and T_1-T_2Q is again all-pass. If, however,

$$|T_1(\infty)| = \|T_1\|_\infty ,$$

then $Q=0$ is also optimal, but $T_1 - T_2 Q$ is not all-pass unless T_1 happens to be.

Example 4.

For an example where an optimal Q doesn't exist take

$$T_1(s) = \frac{1}{s+1}, \quad T_2(s) = \frac{1}{(s+1)^2} .$$

It is claimed that $\alpha=0$. To see this define

$$Q_\epsilon(s) := \frac{s+1}{\epsilon s+1}, \quad \epsilon>0 .$$

Then

$$(T_1 - T_2 Q_\epsilon)(s) = \frac{\epsilon s}{(s+1)(\epsilon s+1)},$$

so from the Bode plot of this function

$$\|T_1 - T_2 Q_\epsilon\|_\infty \leq \epsilon .$$

Thus $\|T_1 - T_2 Q\|_\infty$ can be made arbitrarily small by suitable choice of Q in $\mathbf{RH}_\infty$, i.e. $\alpha=0$. But the only solution of

$$\|T_1 - T_2 Q\|_\infty = 0$$

is $Q(s)=s+1$, which doesn't belong to $\mathbf{RH}_\infty$. So an optimal Q doesn't exist.

The following theorem provides a sufficient condition for an optimal Q to exist.

Theorem 1. An optimal Q exists if the ranks of the two matrices $T_2(j\omega)$ and $T_3(j\omega)$ are constant for all $0 \leq \omega \leq \infty$.

The proof of this theorem involves some advanced tools from functional analysis and so is omitted. The rank conditions will be *assumed* to hold for the remainder of this chapter. Note that they don't hold in Examples 3 and 4; since an optimal Q does exist in Example 3, the conditions are not necessary for existence.

Let's see what the rank conditions mean for a specific control problem.

Example 5.

Consider the tracking problem outlined in Chapter 3. We have

$$G = \begin{bmatrix} G_{11} & G_{12} \\ G_{21} & G_{22} \end{bmatrix}$$

$$G_{11} = \begin{bmatrix} W \\ 0 \end{bmatrix}, \quad G_{12} = \begin{bmatrix} -P \\ \rho I \end{bmatrix}$$

$$G_{21} = \begin{bmatrix} W \\ 0 \end{bmatrix}, \quad G_{22} = \begin{bmatrix} 0 \\ P \end{bmatrix}.$$

To conform with the assumptions of Chapter 4 (G proper, G_{22} strictly proper, G stabilizable), assume $W \in \mathbf{RH}_\infty$ and P is strictly proper.

Bring in a doubly-coprime factorization of P:

$$P = N_P M_P^{-1} = \tilde{M}_P^{-1} \tilde{N}_P$$

$$\begin{bmatrix} \tilde{X}_P & -\tilde{Y}_P \\ -\tilde{N}_P & \tilde{M}_P \end{bmatrix} \begin{bmatrix} M_P & Y_P \\ N_P & X_P \end{bmatrix} = I .$$

Then a doubly-coprime factorization of G_{22} as in (4.5.1) is

$$N_2 = \begin{bmatrix} 0 \\ N_P \end{bmatrix}, \quad M_2 = M_P$$

$$\tilde{N}_2 = \begin{bmatrix} 0 \\ \tilde{N}_P \end{bmatrix}, \quad \tilde{M}_2 = \begin{bmatrix} I & 0 \\ 0 & \tilde{M}_P \end{bmatrix}$$

$$X_2 = \begin{bmatrix} I & 0 \\ 0 & X_P \end{bmatrix}, \quad Y_2 = [0 \quad Y_P]$$

$$\tilde{X}_2 = \tilde{X}_P \,, \quad \tilde{Y}_2 = [0 \quad \tilde{Y}_P] \,.$$

From (4.5.3)

$$T_2 = G_{12} M_2 = \begin{bmatrix} -N_P \\ \rho M_P \end{bmatrix}$$

$$T_3 = \tilde{M}_2 G_{21} = \begin{bmatrix} W \\ 0 \end{bmatrix} .$$

If $\rho > 0$, then rank $T_2(j\omega)$ is constant for all $0 \leq \omega \leq \infty$; this follows from right-coprimeness of N_P, M_P. If ρ were equal to zero, then $T_2(\infty)$ would equal 0 because P is strictly proper. Thus T_2 satisfies the rank condition iff the control energy is weighted in the performance function (see equation (3.1)).

The matrix T_3 satisfies the rank condition iff rank $W(j\omega)$ is constant for all $0 \leq \omega \leq \infty$. This is a type of nonsingularity condition on the reference signal r in Figure 4 of Chapter 3.

Exercise 1. Find

$$\inf \{ \| T_1 - T_2 Q \|_\infty : Q \in \mathbf{RH}_\infty \}$$

for

$$T_1(s) = \frac{10s+1}{s+1}, \quad T_2(s) = \frac{s}{s+1} .$$

6.2 Solution in the Scalar-Valued Case

This section contains a complete solution to the model-matching problem when the T_i 's are scalar-valued. As mentioned in the previous section, we may assume $T_3 = 1$. To conform with Theorem 1.1 it is also assumed that $T_2(j\omega) \neq 0$ for all $0 \leq \omega \leq \infty$. Finally, it is assumed that $T_2^{-1} \notin \mathbf{RH}_\infty$ to avoid the trivial instance of the problem.

We begin with the notions of inner and outer functions. A scalar-valued function T in $\mathbf{RH}_\infty$ is *inner* if $T^\sim T = 1$, i.e.

$$T(-s)T(s) = 1 ,$$

and *outer* if it has no zeros in Re $s > 0$. Examples of inner functions are

$$1, \quad \frac{1-s}{1+s}, \quad \frac{1-s+s^2}{1+s+s^2} .$$

Inner functions have pole-zero symmetry with respect to the imaginary axis:

$s = s_0$ is a zero iff its mirror image $s = -\bar{s}_0$ is a pole. Observe that the zeros of an inner function all lie inside the right half-plane (hence the adjective "inner"). Examples of outer functions are

$$1\,,\ \frac{s+2}{s+1}\,,\ \frac{s}{s+1}\,.$$

The zeros of an outer function all lie outside the right half-plane Re $s > 0$ (hence "outer"). In electrical engineering terminology, an inner function is stable and all-pass with unit magnitude and an outer function is stable and minimum phase.

Lemma 1. Every scalar-valued function T in $\mathbf{RH}_\infty$ has a factorization $T = T_i T_o$ with T_i inner and T_o outer. If $T(j\omega) \neq 0$ for all $0 \leq \omega \leq \infty$, then $T_o^{-1} \in \mathbf{RH}_\infty$.

Proof. Let T_i be the product of all factors of the form

$$\frac{a - s}{\bar{a} + s}$$

where a ranges over all zeros of T in Re $s > 0$, counting multiplicities, and define $T_o := T / T_i$. Then T_i and T_o are inner and outer respectively, and $T = T_i T_o$. If T is not strictly proper and has no zeros on the imaginary axis, then T_o has these two properties too, so $T_o^{-1} \in \mathbf{RH}_\infty$. □

A factorization of the above form is called an *inner-outer factorization.* (A state-space procedure for doing matrix inner-outer factorization will be developed in Section 7.4.)

Exercise 1. Prove that inner and outer factors are unique up to sign, i.e. if

$$U_i U_o = V_i V_o$$

$U_i\,,\ V_i$ inner

$U_o\,,\ V_o$ outer and nonzero ,

then either

$$U_i = V_i\,,\ U_o = V_o$$

or

$$U_i = -V_i\,,\; U_o = -V_o\;.$$

Returning to the model-matching problem, bring in an inner-outer factorization of T_2: $T_2 = T_{2i}T_{2o}$. For Q in $\mathbf{RH}_\infty$ we have

$$\begin{aligned} \|T_1 - T_2 Q\|_\infty &= \|T_1 - T_{2i}T_{2o}Q\|_\infty \\ &= \|T_{2i}(T_{2i}^{-1}T_1 - T_{2o}Q)\|_\infty \\ &= \|T_{2i}^{-1}T_1 - T_{2o}Q\|_\infty \qquad (1) \\ &= \|R - X\|_\infty\,, \end{aligned}$$

where

$$R := T_{2i}^{-1}T_1 \qquad (2)$$

$$X := T_{2o}Q\;. \qquad (3)$$

Equality in (1) follows from the property $|T_{2i}(j\omega)| = 1$. Note that $R \in \mathbf{RL}_\infty$. Also, since T_{2o} and $T_{2o}^{-1} \in \mathbf{RH}_\infty$, (3) sets up a one-to-one correspondence between functions Q in $\mathbf{RH}_\infty$ and functions X in $\mathbf{RH}_\infty$. We conclude that

$$\alpha = \inf\{\|R - X\|_\infty : X \in \mathbf{RH}_\infty\} \qquad (4a)$$

$$= \mathrm{dist}(R, \mathbf{RH}_\infty)\;. \qquad (4b)$$

A function X in $\mathbf{RH}_\infty$ satisfying

$$\alpha = \|R - X\|_\infty$$

will be called *optimal.* An optimal X yields an optimal Q via (3).

The optimization problem in (4) is much like that in Section 5.2, finding the distance from R to $\mathbf{H}_\infty$. Since $\mathbf{RH}_\infty \subset \mathbf{H}_\infty$ we have

$$\mathrm{dist}(R, \mathbf{RH}_\infty) \geq \mathrm{dist}(R, \mathbf{H}_\infty)\;. \qquad (5)$$

It will turn out that the distances are in fact equal because R is real-rational, so that, from (4) and Nehari's theorem, α equals $\|\Gamma_R\|$.

For the main results we shall use the machinery of Chapter 5. Factor R as

$$R = R_1 + R_2$$

with R_1 strictly proper and analytic in Re $s \leq 0$ and R_2 in $\mathbf{RH}_\infty$. Then R_1 has a minimal state-space realization

$$R_1(s) = [A, B, C, 0] \tag{6}$$

with A antistable. The controllability and observability grammians are the unique solutions of

$$AL_c + L_c A^T = BB^T \tag{7}$$

$$A^T L_o + L_o A = C^T C . \tag{8}$$

Let λ^2 equal the largest eigenvalue of $L_c L_o$ and let w be a corresponding eigenvector:

$$L_c L_o w = \lambda^2 w . \tag{9}$$

Defining $v := \lambda^{-1} L_o w$, we have the pair of equations

$$L_c v = \lambda w \tag{10}$$

$$L_o w = \lambda v . \tag{11}$$

Finally, define the real-rational functions

$$f(s) := [A, w, C, 0] \tag{12}$$

$$g(s) := [-A^T, v, B^T, 0] . \tag{13}$$

Observe that $f \in \mathbf{RH}_2^\perp$ and $g \in \mathbf{RH}_2$.

Lemma 2. The functions f and g satisfy the equations

$$\Gamma_R g = \lambda f \tag{14}$$

$$\Gamma_R^* f = \lambda g . \tag{15}$$

Proof. To prove (14) start with (7). Add and subtract sL_c on the left-hand side to get

$$-(s-A)L_c + L_c(s+A^T) = BB^T .$$

Now pre-multiply by $C(s-A)^{-1}$ and post-multiply by $(s+A^T)^{-1}v$ to get

$$-CL_c(s+A^T)^{-1}v + C(s-A)^{-1}L_c v$$

$$= C(s-A)^{-1}BB^T(s+A^T)^{-1}v \ . \tag{16}$$

The first function on the left-hand side belongs to $\mathbf{H}_2$; from (10) and (12) the second function equals $\lambda f(s)$; and from (6) and (13) the function on the right-hand side equals $R_1(s)g(s)$. Project both sides of (16) onto $\mathbf{H}_2^\perp$ to get

$$\lambda f = \Pi_1 R_1 g$$

$$= \Gamma_{R_1} g \ .$$

But $\Gamma_{R_1} = \Gamma_R$; hence (14) holds.

Equation (15) is proved similarly starting with (8). □

Vectors f and g satisfying (14) and (15) are said to be a Schmidt pair for the operator Γ_R.

Exercise 2. Prove that $f^\sim f = g^\sim g$.

Notice from (14) and (15) that

$$\Gamma_R^* \Gamma_R g = \lambda^2 g \ , \tag{17}$$

i.e. g is an eigenvector of $\Gamma_R^* \Gamma_R$ corresponding to λ^2, the largest eigenvalue (by Theorem 5.1.3).

Theorem 1. The infimal model-matching error α equals $\|\Gamma_R\|$, the unique optimal X equals $R - \alpha f/g$, and, for the optimal Q, $T_1 - T_2 Q$ is all-pass.

Proof. From Nehari's theorem there exists a function X in $\mathbf{H}_\infty$ such that

$$\|R - X\|_\infty = \|\Gamma_R\| \ . \tag{18}$$

It is claimed that

$$(R - X)g = \Gamma_R g \ . \tag{19}$$

To prove this, define $h := (R - X)g$ and look at the $\mathbf{L}_2$-norm of $h - \Gamma_R g$:

$$\|h - \Gamma_R g\|_2^2 = \langle h - \Gamma_R g, h - \Gamma_R g \rangle$$

$$= \langle h, h \rangle + \langle \Gamma_R g, \Gamma_R g \rangle$$

$$- \langle h, \Gamma_R g \rangle - \langle \Gamma_R g, h \rangle . \qquad (20)$$

Being in $\mathbf{H}_2^\perp$, $\Gamma_R g$ is orthogonal to the $\mathbf{H}_2$-component of h. Thus

$$\langle h, \Gamma_R g \rangle = \langle \Pi_1 h, \Gamma_R g \rangle .$$

But

$$\Pi_1 h = \Pi_1 (R - X) g$$

$$= \Pi_1 R g$$

$$= \Gamma_R g .$$

Hence

$$\langle h, \Gamma_R g \rangle = \langle \Gamma_R g, \Gamma_R g \rangle . \qquad (21)$$

Using (21) in (20) we get

$$\|h - \Gamma_R g\|_2^2 = \langle h, h \rangle - \langle \Gamma_R g, \Gamma_R g \rangle$$

$$= \langle h, h \rangle - \langle g, \Gamma_R^* \Gamma_R g \rangle$$

$$= \langle h, h \rangle - \lambda^2 \langle g, g \rangle \quad \text{from (17)}$$

$$= \|(R - X) g\|_2^2 - \lambda^2 \|g\|_2^2$$

$$\leq (\|R - X\|_\infty^2 - \lambda^2) \|g\|_2^2 .$$

But $\|R - X\|_\infty = \lambda$ from (18), so $h = \Gamma_R g$. This proves the claim.

From the claim we get the first two assertions in the theorem statement as follows. Every X in $\mathbf{H}_\infty$ satisfying (18) also satisfies (19). But (19) has a unique solution, namely,

$$X = R - \lambda f / g .$$

(Here (14) is used.) Since the latter function belongs to $\mathbf{RH}_\infty$, equality holds in (5). Thus from (4) and Nehari's theorem

$$\alpha = \lambda = \|\Gamma_R\| .$$

Therefore $X = R - \alpha f / g$.

To prove the third assertion in the theorem statement start with (2) and (3) to get

$$T_1 - T_2 Q = T_{2i} (R - X) .$$

To prove that $T_1 - T_2 Q$ is all-pass, it suffices to show that $R - X$ is all-pass. But $R - X = \lambda f / g$ and

$$| f(j\omega) | = | g(j\omega) |$$

from the preceding exercise. □

Let's summarize this section in the form of an algorithm to compute α and the optimal Q .

Step 1. Do an inner-outer factorization

$$T_2 = T_{2i} T_{2o} .$$

Step 2. Define

$$R := T_{2i}^{-1} T_1$$

and find a minimal realization

$$R(s) = [A, B, C, 0] + (\text{a function in } \mathbf{RH}_\infty).$$

Step 3. Solve the equations

$$AL_c + L_c A^T = BB^T$$

$$A^T L_o + L_o A = C^T C .$$

Step 4. Find the maximum eigenvalue λ^2 of $L_c L_o$ and a corresponding eigenvector w .

Step 5. Define

$$f(s) = [A, w, C, 0]$$

$$g(s) = [-A^T, \lambda^{-1} L_o w, B^T, 0]$$

$$X = R - \lambda f / g .$$

Step 6. Set $\alpha = \lambda$ and $Q = T_{2o}^{-1} X$.

Example 1.

The algorithm applied to

$$T_1(s) = \frac{s+1}{10s+1}, \quad T_2(s) = \frac{(s-1)(s-5)}{(s+2)^2}$$

goes like this:

Step 1.

$$T_{2i}(s) = \frac{(s-1)(s-5)}{(s+1)(s+5)}, \quad T_{2o}(s) = \frac{(s+1)(s+5)}{(s+2)^2}$$

Step 2.

$$R(s) = \frac{(s+1)^2(s+5)}{(10s+1)(s-1)(s-5)}$$

$$A = \begin{bmatrix} 1 & 0 \\ 0 & 5 \end{bmatrix}, \quad B = \begin{bmatrix} -6/11 \\ 90/51 \end{bmatrix}$$

$$C = [1 \ \ 1]$$

Step 3.

$$L_c = \begin{bmatrix} .1488 & -.1604 \\ -.1604 & .3114 \end{bmatrix}, \quad L_o = \begin{bmatrix} .5 & .1667 \\ .1667 & .1 \end{bmatrix}$$

Step 4.

$$\lambda = .2021, \quad w = \begin{bmatrix} 1 \\ -.7769 \end{bmatrix}$$

Step 5.

$$f(s) = \frac{.2231s - 4.223}{(s-1)(s-5)}, \quad g(s) = \frac{-.2231s - 4.2231}{(s+1)(s+5)}$$

$$X(s) = 3.021 \frac{(s+1)(s+5)}{(10s+1)(s+18.93)}$$

Step 6.

$$\alpha = .2021, \quad Q(s) = 3.021 \frac{(s+2)^2}{(10s+1)(s+18.93)}$$

6.3 A Single-Input, Single-Output Design Example

This section illustrates an application of the previous one to a very simple design problem. It is emphasized that the design problem is not meant to be entirely realistic; in particular, it lacks bandwidth constraints on the controller and ignores stability margin.

Consider the single-loop feedback system in Figure 1 and assume P is strictly proper and K is proper. The transfer function from reference input, w, to tracking error, z, is the *sensitivity function*

$$S := 1/(1+PK). \tag{1}$$

Suppose w has its energy concentrated on a frequency band, say $[0,\omega_1]$. Then an appropriate performance specification would be

$$|S(j\omega)| < \epsilon \text{ for all } \omega \text{ in } [0,\omega_1],$$

where ϵ is pre-specified and less than 1. For example, $\epsilon=.01$ would provide less than 1% tracking error of all sinusoids in the frequency range $[0,\omega_1]$. To simplify notation slightly, define the $\mathbf{L}_\infty$-function

$$\chi(j\omega) := 1 \text{ if } |\omega| \le \omega_1$$

$$:= 0 \text{ if } |\omega| > \omega_1 .$$

Then the performance specification is

$$\|\chi S\|_\infty < \epsilon . \tag{2}$$

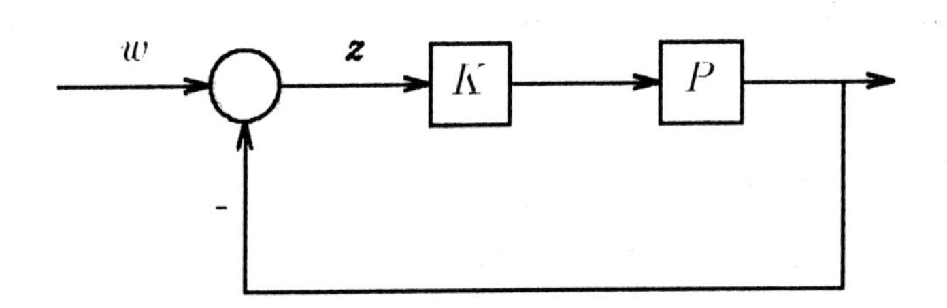

Figure 6.3.1. Single-loop feedback system

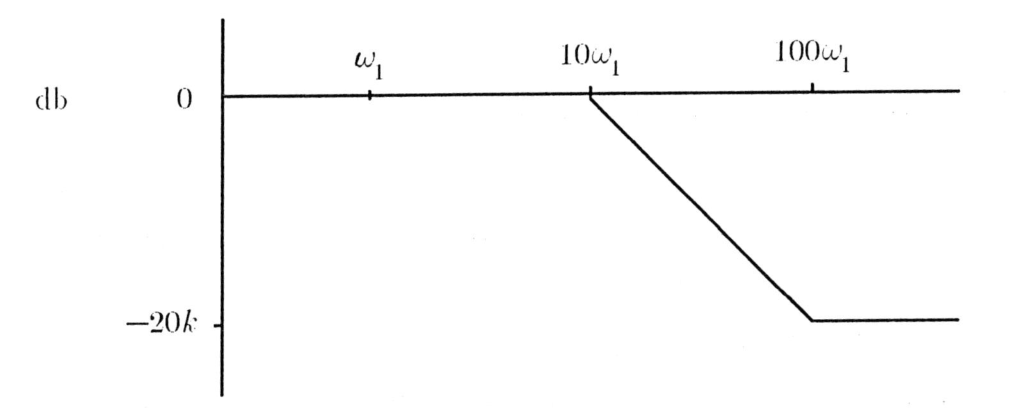

Figure 6.3.2. Bode plot of W^{-k}

We consider the following *design problem*: given P, ω_1, and ϵ, design K to achieve specification (2) subject to the constraint of feedback stability.

Figure 1 looks like Figure 4.4.1 with $G := -P$. Thus Theorem 4.4.1 yields a formula for all stabilizing controllers. Bring in a coprime factorization of $-P$:

$$-P = N/M \tag{3}$$

$$MX - NY = 1 . \tag{4}$$

As usual, N, M, X, Y all $\in \mathbf{RH}_\infty$. Then all proper stabilizing K's are given by the formula

$$K = (Y - MQ)/(X - NQ) \tag{5}$$

$$Q \in \mathbf{RH}_\infty .$$

Substitute (3) and (5) into (1) and simplify using (4) to get

$$S = MX - MNQ . \tag{6}$$

So the design problem is reduced to determining a function Q in $\mathbf{RH}_\infty$ such that

$$\|\chi(MX - MNQ)\|_\infty < \epsilon . \tag{7}$$

The left-hand side of (7) has the form

$$\|T_1 - T_2 Q\|_\infty ,$$

but the T_i's belong to $\mathbf{L}_\infty$ instead of $\mathbf{RH}_\infty$ as required in the previous section. So let's approximate χ by an $\mathbf{RH}_\infty$-function W^k where

$$W(s) := (.01\omega_1^{-1}s + 1)/(.1\omega_1^{-1}s + 1) .$$

The piecewise-linear magnitude Bode plot of W^k is shown in Figure 2. Its characteristics are that it is nearly unity on the operating band $[0, \omega_1]$ (in fact to a decade above) and then it drops off to -20k db at high frequency. The integer k remains unspecified at this stage; it's regarded as a design parameter and its function is explained below.

We now observe that if $\|W^k S\|_\infty < \epsilon$, then $\|\chi S\|_\infty$ is less than ϵ, or at least not much larger than ϵ. This is because $|W(j\omega)|^k$ and $\chi(j\omega)$ are nearly equal on the operating band. Using (6) again we conclude that the design problem reduces to this: find $k \geq 1$ and Q in $\mathbf{RH}_\infty$ such that

$$\| W^k (MX - MNQ) \|_\infty < \epsilon ,$$

i.e.

$$\| T_1 - T_2 Q \|_\infty < \epsilon , \tag{8}$$

where $T_1 := W^k MX$, $T_2 := W^k MN$. Clearly T_1 and $T_2 \in \mathbf{RH}_\infty$.

As in (1.1) define

$$\alpha := \inf \{ \| T_1 - T_2 Q \|_\infty : Q \in \mathbf{RH}_\infty \} .$$

Here α depends on the integer k, so let's write α_k. The following result justifies the introduction of k.

Lemma 1. If P has no zeros on the imaginary axis in the frequency range $[0, 100\omega_1]$, then

$$\lim_{k \to \infty} \alpha_k = 0 .$$

Proof. Let $\chi_1(j\omega)$ equal 1 up to $\omega = 100\omega_1$ and 0 beyond. We have

$$\alpha_k = \inf \{ \| W^k M (X - NQ) \|_\infty : Q \in \mathbf{RH}_\infty \} .$$

For an arbitrary $\delta > 0$ choose a function Q in $\mathbf{RH}_\infty$ such that

$$\| \chi_1 (X - NQ) \|_\infty < \delta . \tag{9}$$

It's possible to do this because N has no zeros on the imaginary axis in the frequency range $[0, 100\omega_1]$; so Q can be chosen to approximate X/N over this segment of the imaginary axis. (A rigorous justification of Q's existence uses Runge's theorem.) Now we get

$$\begin{aligned} \alpha_k &\leq \| W^k M (X - NQ) \|_\infty \\ &\leq \| M \|_\infty \max(\| \chi_1 W^k (X - NQ) \|_\infty , \| (1 - \chi_1) W^k (X - NQ) \|_\infty) . \end{aligned} \tag{10}$$

Since $\| W^k \|_\infty \leq 1$ we have from (9)

$$\| \chi_1 W^k (X - NQ) \|_\infty < \delta . \tag{11}$$

Also

$$\|(1-\chi_1)W^k(X-NQ)\|_\infty \le \|X-NQ\|_\infty \|(1-\chi_1)W^k\|_\infty . \tag{12}$$

Using (11) and (12) in (10) gives

$$\alpha_k \le \|M\|_\infty \max(\delta, \|X-NQ\|_\infty \|(1-\chi_1)W^k\|_\infty) .$$

It follows from Figure 2 that

$$\lim_{k \to \infty} \|(1-\chi_1)W^k\|_\infty = 0 .$$

Thus for sufficiently large k, $\alpha_k \le \|M\|_\infty \delta$. Since δ was arbitrary, we conclude that $\alpha_k \to 0$. □

In view of Lemma 1 the design problem reduces to choosing k such that $\alpha_k < \epsilon$ and then finding a Q in $\mathbf{RH}_\infty$ to satisfy (8). There remains one hitch though: it's not true that $T_2(j\omega) \ne 0$ for all $0 \le \omega \le \infty$; in particular, T_2 is strictly proper because P is. So in fact the infimum

$$\inf \{\|T_1 - T_2 Q\|_\infty : Q \in \mathbf{RH}_\infty\} \tag{13}$$

is not achieved.

Let's see how to approach the optimization problem (13) when P has neither poles nor zeros on the imaginary axis. Then $T_2(j\omega) \ne 0$ for all $0 \le \omega < \infty$, but $T_2(j\infty) = 0$. Introduce a polynomial

$$V(s) = (s+1)^l$$

where the integer l equals the *relative degree* of T_2, number of poles minus number of zeros. Then $T_2 V$ is proper but not strictly proper. Instead of (13), consider

$$\inf \{\|T_1 - T_2 V Q_1\|_\infty : Q_1 \in \mathbf{RH}_\infty\} .$$

The theory of the previous section applies to this problem, and there exists a unique optimal Q_1. Now define $Q := VQ_1$. Then Q is stable but not proper. We can approximate Q on the operating band $[0,\omega_1]$ by rolling off at high frequency, say a decade above ω_1:

$$Q_a(s) := Q(s)/(.1\omega_1^{-1}s+1)^l .$$

Let's recap by listing the steps in the design procedure. The input data are P, ω_1, and ϵ, and P has neither poles nor zeros on the imaginary axis.

Step 1. Do a coprime factorization of $-P$:

$$-P = N/M$$

$$MX - NY = 1.$$

Step 2. Define the weighting function

$$W(s) = (.01\omega_1^{-1}s+1)/(.1\omega_1^{-1}s+1)$$

and initialize k to 1.

Step 3. Set

$$T_1 = W^k MX$$

$$T_2 = W^k MN$$

$$V(s) = (s+1)^l$$

$$l = \text{relative degree of } P .$$

Step 4. By the method of Section 2 compute

$$\alpha_k = \min\{\|T_1 - T_2 V Q_1\|_\infty : Q_1 \in \mathbf{RH}_\infty\} .$$

If $\alpha_k \geq \epsilon$, increment k by 1 and go back to Step 3. Otherwise, continue.

Step 5. By the method of Section 2 compute the function Q_1 in $\mathbf{RH}_\infty$ such that

$$\alpha_k = \|T_1 - T_2 V Q_1\|_\infty .$$

Step 6. Define

$$Q_a(s) = V(s)Q_1(s)/(.1\omega_1^{-1}s+1)^l$$

$$K = (Y - MQ_a)/(X - NQ_a) .$$

How the performance specification (2) is (approximately) achieved is summarized as follows. We have in succession

$$\begin{aligned} ||\chi S||_\infty &= ||\chi/(1+PK)||_\infty \\ &= ||\chi(MX-MNQ_a)||_\infty \\ &\approx ||\chi(MX-MNQ_1)||_\infty \\ &< \epsilon . \end{aligned}$$

Example 1.

This example illustrates the above procedure for the nonminimum phase plant

$$P(s) = \frac{(s-1)(s-2)}{(s+1)(s^2+s+1)}$$

and the performance specs ω_1=0.01, ϵ=0.1 (-20 db). Thus we are to achieve less than 10% tracking error up to 0.01 rad/s.

Step 1.

$$N = -P, \quad M = 1, \quad X = 1, \quad Y = 0$$

Step 2.

$$W(s) = \frac{s+1}{10s+1}$$

Step 3.

$$T_1(s) = \left(\frac{s+1}{10s+1}\right)^k$$

$$T_2(s) = -\left(\frac{s+1}{10s+1}\right)^k \frac{(s-1)(s-2)}{(s+1)(s^2+s+1)}$$

$$V(s) = s+1$$

Step 4.

$$\alpha_1 = .2299$$

$$\alpha_2 = .05114$$

Thus $\alpha_2 < \epsilon$.

Step 5.

$$Q_1(s) = -6.114 \frac{(s+.3613)(s^2+s+1)}{(s+4.656)(s+1)^2}$$

Step 6.

$$Q_a(s) = -6.114 \frac{(s+.3613)(s^2+s+1)}{(s+4.656)(s+1)(10s+1)}$$

$$K(s) = .6114 \frac{(s+.3613)(s+1)(s^2+s+1)}{(s+.004698)(s+.5280)(s^2+5.612s+9.599)}$$

The Bode magnitude plot for this design is shown in Figure 3. The design didn't quite meet the spec ($|S(.01j)| = -18.4$ db); this is because of the approximations made during the procedure.

Notes and References

The question of when the model-matching problem has a solution is dealt with in, for example, Francis (1983).

An example of the scalar-valued model-matching problem is the weighted sensitivity problem. This was solved in Zames and Francis (1983) and Francis and Zames (1984). Lemma 2.2 is from Glover (1984) and the proof of Theorem 2.1 is adapted from the elegant paper of Sarason (1967). The nice state-space formulas for f and g (equations (2.12) and (2.13)) are due to Silverman and Bettayeb (1980).

The design problem of Section 3 can be made more realistic by incorporating global bounds on $|S|$, as for example in O'Young and Francis (1985), or bounds on the complementary sensitivity function $1-S$, as in Foo and Postlethwaite (1984), Kwakernaak (1985), and Verma and Jonckheere (1984).

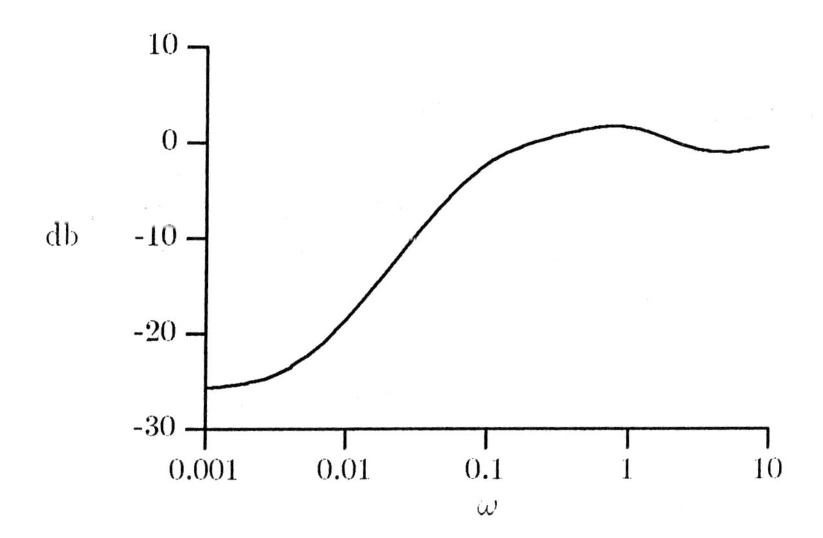

Figure 6.3.3. Magnitude Bode plot of S, Example 1

CHAPTER 7

FACTORIZATION THEORY

The purpose of this chapter is to develop some basic tools required for the solution of the model-matching problem in the matrix-valued case.

7.1 The Canonical Factorization Theorem

Consider a scalar-valued real-rational function $G(s)$ which is proper and has no poles on the imaginary axis; thus $G \in \mathbf{RL}_\infty$. Suppose in addition that G has no zeros on the imaginary axis nor at infinity. Then $G^{-1} \in \mathbf{RL}_\infty$ too. Now consider the problem of factoring G as $G = G_+ G_-$ where G_+ has all its poles and zeros in Re $s > 0$ and G_- has all its poles and zeros in Re $s < 0$. Furthermore, we require that G_+ and G_- be proper and have proper inverses. When does such a factorization exist? A moment's thought will lead to the conclusion that G has such a factorization iff it has the property

$$\{\text{no. poles in Re } s < 0\} = \{\text{no. zeros in Re } s < 0\},$$

or equivalently

$$\{\text{no. poles}\} = \{\text{no. zeros in Re } s < 0\} + \{\text{no. poles in Re } s > 0\}.$$

The purpose of this section is to derive the analogous condition in the matrix case and give a procedure for doing such a factorization.

Let $G(s)$ be a square matrix such that G, $G^{-1} \in \mathbf{RL}_\infty$. Thus G and its inverse are proper and have no poles on the imaginary axis. Our goal is to factor G as $G = G_+ G_-$, where the factors G_+ and G_- are square and have the properties

$$G_-,\ G_-^{-1} \in \mathbf{RH}_\infty$$

$$G_+^\sim,\ (G_+^{-1})^\sim \in \mathbf{RH}_\infty .$$

The latter condition means that G_+ and its inverse are proper and analytic in Re $s<0$. For ease of reference let's call a factorization as just described a *canonical factorization* of G.

We begin with a minimal realization,

$$G(s)=[A, B, C, D].$$

Since $G(\infty)=D$ and $G^{-1}\in\mathbf{RL}_\infty$, we see that D is invertible. Define

$$A^\times := A - BD^{-1}C$$

and write the state-space equations for G:

$$\dot{x} = Ax + Bu$$

$$y = Cx + Du .$$

Re-arrange to get y as input and u as output:

$$\dot{x} = A^\times x + BD^{-1}y$$

$$u = -D^{-1}Cx + D^{-1}y .$$

Thus

$$G(s)^{-1} = [A^\times, BD^{-1}, -D^{-1}C, D^{-1}]. \tag{1}$$

Next we recall the notions of modal subspaces. Suppose A is of dimension $n\times n$. Let $\alpha(s)$ denote the characteristic polynomial of A and factor it as $\alpha(s)=\alpha_-(s)\alpha_+(s)$, where α_- has all its zeros in Re $s<0$ and α_+ has all its zeros in Re $s>0$. (There are no zeros on the imaginary axis.) Then the *modal subspaces* of $\mathbf{R}^n$ relative to A are

$$\mathbf{X}_-(A) := \text{Ker } \alpha_-(A)$$

$$\mathbf{X}_+(A) := \text{Ker } \alpha_+(A),$$

where Ker denotes kernel (null space). It can be shown that $\mathbf{X}_-(A)$ is spanned by the generalized (real) eigenvectors of A corresponding to eigenvalues in Re $s<0$; similarly for $\mathbf{X}_+(A)$. These two modal subspaces are *complementary,* i.e. they're independent and their sum is all of $\mathbf{R}^n$. Thus we write

$$\mathbf{R}^n = \mathbf{X}_-(A) \oplus \mathbf{X}_+(A).$$

(The two subspaces are not orthogonal in general.)

Bases for modal subspaces can be computed using standard numerical linear algebra. For example, suppose it's desired to compute a basis for $\mathbf{X}_-(A)$. Transform A to real Schur form, ordering the eigenvalues with increasing real part. Then partition the Schur form as

$$\begin{bmatrix} A_1 & A_2 \\ 0 & A_4 \end{bmatrix}, \tag{2}$$

where A_1 has all its eigenvalues in Re $s < 0$ and A_4 has all its eigenvalues in Re $s \geq 0$. Let T be the orthogonal transformation matrix, i.e. $T^T A T$ equals matrix (2), and partition T conformably:

$$T = [T_1 \quad T_2] .$$

Then $\mathbf{X}_-(A) = \text{Im } T_1$, the column span of T_1.

Now consider the two modal subspaces $\mathbf{X}_-(A^\times)$ and $\mathbf{X}_+(A)$. The former is associated with left half-plane zeros of G and the latter with right half-plane poles of G.

Theorem 1. G has a canonical factorization if $\mathbf{X}_-(A^\times)$ and $\mathbf{X}_+(A)$ are complementary.

Since the proof is constructive and is used several times in the sequel, it's useful to present it as an algorithm.

Step 1. Obtain real matrices T_1 and T_2, each with full column rank, such that

$$\mathbf{X}_-(A^\times) = \text{Im } T_1 \tag{3}$$

$$\mathbf{X}_+(A) = \text{Im } T_2 . \tag{4}$$

Define $T := [T_1 \quad T_2]$ and note that T is square and nonsingular by the hypothesis of the theorem.

Step 2. Introduce the partitions

$$T^{-1}AT = \begin{bmatrix} A_1 & A_2 \\ A_3 & A_4 \end{bmatrix} \tag{5}$$

$$T^{-1}B = \begin{bmatrix} B_1 \\ B_2 \end{bmatrix} \tag{6}$$

$$CT = [C_1 \quad C_2] \tag{7}$$

corresponding to the above partition of T, e.g. A_1 is square and its dimension equals that of $\mathbf{X}_-(A^\times)$.

Step 3. Define

$$G_+(s) := [A_4, B_2, C_2, D] \tag{8}$$

$$G_-(s) := [A_1, B_1, D^{-1}C_1, I]\ . \tag{9}$$

The proof that these steps do indeed result in a canonical factorization uses the following two exercises.

Exercise 1. In (5) show that $A_2 = 0$, A_1 is stable, and A_4 is antistable (all eigenvalues in Re $s > 0$). (Hint: use (4).)

From the partitions (5)-(7) we have

$$T^{-1}A^\times T = \begin{bmatrix} A_1 - B_1D^{-1}C_1 & -B_1D^{-1}C_2 \\ A_3 - B_2D^{-1}C_1 & A_4 - B_2D^{-1}C_2 \end{bmatrix}.$$

Exercise 2. Show that $A_3 - B_2D^{-1}C_1 = 0$, $A_1 - B_1D^{-1}C_1$ is stable, and $A_4 - B_2D^{-1}C_2$ is antistable. (Hint: use (3).)

It follows from the first exercise that $G_+^\sim$, $G_- \in \mathbf{RH}_\infty$. Also, since

$$G_+(s)^{-1} = [A_4 - B_2D^{-1}C_2, B_2D^{-1}, -D^{-1}C_2, D^{-1}]$$

$$G_-(s)^{-1} = [A_1 - B_1D^{-1}C_1, B_1, -D^{-1}C_1, I]\ ,$$

we conclude from the second exercise that G_-^{-1}, $(G_+^{-1})^\sim \in \mathbf{RH}_\infty$. Finally, the verification that $G = G_+G_-$ is as follows:

$$G_+(s)G_-(s) = [A_4, B_2, C_2, D] \times [A_1, B_1, D^{-1}C_1, I]$$

$$= \left[\begin{bmatrix} A_1 & 0 \\ B_2 D^{-1} C_1 & A_4 \end{bmatrix}, \begin{bmatrix} B_1 \\ B_2 \end{bmatrix}, [C_1 \ \ C_2] , D \right]$$

$$= \left[\begin{bmatrix} A_1 & 0 \\ A_3 & A_4 \end{bmatrix}, \begin{bmatrix} B_1 \\ B_2 \end{bmatrix}, [C_1 \ \ C_2] , D \right]$$

$$= [T^{-1}AT, T^{-1}B, CT, D]$$

$$= [A, B, C, D]$$

$$= G(s) .$$

The converse of Theorem 1 is true too, but we shan't need it.

7.2 The Hamiltonian Matrix

In this section we examine the modal subspaces relative to a certain type of matrix, namely, a *Hamiltonian matrix* of the form

$$H := \begin{bmatrix} A & -R \\ -Q & -A^T \end{bmatrix} . \tag{1}$$

Here A , Q , and R are real $n \times n$ matrices, Q and R are symmetric, and R is either positive semi-definite or negative semi-definite. The modal subspaces of H live in $\mathbf{R}^{2n}$.

We consider first the simpler case where $R = 0$ and A is stable. The Lyapunov equation

$$A^T X + XA + Q = 0 \tag{2}$$

has a unique solution X .

Theorem 1. ($R=0$, A stable) The modal subspaces relative to H are

$$\mathbf{X}_+(H) = \operatorname{Im}\begin{bmatrix} 0 \\ I \end{bmatrix}$$

$$\mathbf{X}_-(H) = \operatorname{Im}\begin{bmatrix} I \\ X \end{bmatrix}.$$

Proof. Define the nonsingular matrix

$$T := \begin{bmatrix} I & 0 \\ X & I \end{bmatrix}.$$

Then

$$T^{-1}HT = \begin{bmatrix} A & 0 \\ 0 & -A^T \end{bmatrix},$$

so

$$\mathbf{X}_+(T^{-1}HT) = \operatorname{Im}\begin{bmatrix} 0 \\ I \end{bmatrix}.$$

Hence

$$\begin{aligned} \mathbf{X}_+(H) &= T\,\mathbf{X}_+(T^{-1}HT) \\ &= \operatorname{Im} T\begin{bmatrix} 0 \\ I \end{bmatrix} \\ &= \operatorname{Im}\begin{bmatrix} 0 \\ I \end{bmatrix}. \end{aligned}$$

Similarly for $\mathbf{X}_-(H)$. □

Now we turn to the general case. It is claimed that the spectrum of H is symmetric with respect to the imaginary axis. To see this, introduce the $2n \times 2n$ matrix

$$J := \begin{bmatrix} 0 & -I \\ I & 0 \end{bmatrix}$$

having the property $J^2 = -I$. Then

$$J^{-1}HJ = -JHJ = -H^T .$$

Thus H and $-H^T$ are similar. Hence λ is an eigenvalue of H iff $-\lambda$ is.

Now assume H has no eigenvalues on the imaginary axis. Then it must have n eigenvalues in Re $s < 0$ and n in Re $s > 0$, i.e. the modal subspaces both have dimension n.

Theorem 2. Assume H has no eigenvalues on the imaginary axis and (A, R) is stabilizable. Then $\mathbf{X}_-(H)$ and Im $\begin{bmatrix} 0 \\ I \end{bmatrix}$ are complementary.

Proof. Introduce a $2n \times n$ real matrix T such that

$$\mathbf{X}_-(H) = \text{Im } T \tag{3}$$

and partition it as

$$T = \begin{bmatrix} T_1 \\ T_2 \end{bmatrix}$$

($n \times n$ blocks). Then $\mathbf{X}_-(H)$ and Im $\begin{bmatrix} 0 \\ I \end{bmatrix}$ are complementary iff T_1 is invertible.

First it is claimed that

$$T_1^T T_2 = T_2^T T_1 . \tag{4}$$

There is a stable $n \times n$ matrix H_- such that

$$HT = TH_- . \tag{5}$$

Again, bring in the $2n \times 2n$ matrix

$$J := \begin{bmatrix} 0 & -I \\ I & 0 \end{bmatrix}$$

and pre-multiply (5) by $T^T J$:

$$T^T JHT = T^T JTH_- . \tag{6}$$

Now, JH is symmetric, hence so is the right-hand side of (6):

$$(T^T JT)H_- = H_-^T T^T J^T T$$

$$= -H_-^T (T^T JT) . \tag{7}$$

Since H_- and $(-H_-^T)$ have disjoint spectra, the unique solution of (7) (a Lyapunov equation) is $T^T JT = 0$. But this is equivalent to (4).

The next claim is that Ker T_1 is invariant under H_-. Let $x \in$Ker T_1. Pre-multiply (5) by $[I \quad 0]$ to get

$$AT_1 - RT_2 = T_1 H_- . \tag{8}$$

Pre-multiply this by $x^T T_2^T$ and post-multiply by x to get

$$-x^T T_2^T R T_2 x = x^T T_2^T T_1 H_- x . \tag{9}$$

But from (4) the right-hand side equals

$$x^T T_1^T T_2 H_- x = 0 .$$

Thus the left-hand side of (9) equals zero, i.e.

$$RT_2 x = 0 . \tag{10}$$

We just used the sign semi-definiteness of R. Post-multiply (8) by x and use (10) to get $T_1 H_- x = 0$, i.e. $H_- x \in$Ker T_1.

Finally, to prove that T_1 is invertible suppose, on the contrary, that Ker $T_1 \neq 0$. Then H_- restricted to Ker T_1 has an eigenvalue λ and a corresponding eigenvector x:

$$H_- x = \lambda x \tag{11}$$

$$\text{Re } \lambda < 0 , \quad 0 \neq x \in \text{Ker } T_1 .$$

Now pre-multiply (5) by $[0 \quad I]$:

$$-QT_1 - A^T T_2 = T_2 H_- . \tag{12}$$

Post-multiply this by x and use (11) to get

$$(A^T + \lambda) T_2 x = 0 . \tag{13}$$

Then (10), (13), and stabilizability of (A, R) imply that $T_2 x = 0$. (Recall that stabilizability implies

$$\text{rank}[A - \mu \quad R] = n$$

for all Re $\mu \geq 0$.) But if $T_1 x = 0$ and $T_2 x = 0$, then $Tx = 0$, which implies $x = 0$, a contradiction. □

The *Riccati equation* associated with H is

$$A^T X + XA - XRX + Q = 0 . \tag{14}$$

Observe that (14) reduces to (2) when $R = 0$. Theorem 2 is related to (14) in the following way.

Corollary 1. Under the same assumptions as in Theorem 2, there exists a unique matrix X such that

$$\mathbf{X}_{-}(H) = \mathrm{Im}\begin{bmatrix} I \\ X \end{bmatrix}. \tag{15}$$

Moreover, X is symmetric, it satisfies (14), and $A - RX$ is stable.

Proof. Continuing with the notation in the previous proof, define $X := T_2 T_1^{-1}$. Then (15) is immediate from (3):

$$\begin{aligned} \mathbf{X}_{-}(H) &= \mathrm{Im}\begin{bmatrix} T_1 \\ T_2 \end{bmatrix} \\ &= \mathrm{Im}\begin{bmatrix} I \\ X \end{bmatrix} T_1 \\ &= \mathrm{Im}\begin{bmatrix} I \\ X \end{bmatrix}. \end{aligned}$$

Uniqueness is easy too, because

$$\mathrm{Im}\begin{bmatrix} I \\ X_1 \end{bmatrix} = \mathrm{Im}\begin{bmatrix} I \\ X_2 \end{bmatrix}$$

iff $X_1 = X_2$.

To prove that X is symmetric, start with

$$XT_1 = T_2 . \tag{16}$$

Pre-multiply by T_1^T to get

$$T_1^T X T_1 = T_1^T T_2 . \tag{17}$$

Now take transpose of (16) and post-multiply by T_1 to get

$$T_1^T X^T T_1 = T_2^T T_1 . \tag{18}$$

Equations (4), (17), and (18) together with nonsingularity of T_1 imply $X^T = X$.

Next, rewrite equation (5) as

$$H \begin{bmatrix} I \\ X \end{bmatrix} T_1 = \begin{bmatrix} I \\ X \end{bmatrix} T_1 H_-$$

i.e.

$$H \begin{bmatrix} I \\ X \end{bmatrix} = \begin{bmatrix} I \\ X \end{bmatrix} T_1 H_- T_1^{-1} . \tag{19}$$

Pre-multiply by $[X \quad -I]$ to get

$$[X \quad -I] H \begin{bmatrix} I \\ X \end{bmatrix} = 0.$$

This is precisely the Riccati equation.

Finally, pre-multiplication of (19) by $[I \quad 0]$ gives

$$A - RX = T_1 H_- T_1^{-1} ,$$

proving that $A - RX$ is stable since H_- is. □

In general the Riccati equation (14) has several solutions, only one of which satisfies (15).

7.3 Spectral Factorization

Consider a square matrix $G(s)$ having the properties

$$G, G^{-1} \in \mathbf{RL}_\infty \tag{1a}$$

$$G^\sim = G \tag{1b}$$

$$G(\infty) > 0 . \tag{1c}$$

Such a matrix has pole and zero symmetry about the imaginary axis. Our goal is to factor G as

$$G = G_-^\sim G_-$$

$$G_-, G_-^{-1} \in \mathbf{RH}_\infty .$$

This is called a *spectral factorization* of G and G_- is a *spectral factor.*

Theorem 1. Every G satisfying (1) has a spectral factorization.

Proof. The proof is constructive. We begin by factoring G as

$$G(s) = D + G_1(s) + G_2(s), \tag{2}$$

where $D := G(\infty)$, G_1 belongs to $\mathbf{RH}_\infty$ and is strictly proper, and $G_2^\sim$ belongs to $\mathbf{RH}_\infty$ and is also strictly proper. It's worth mentioning in passing that such a factorization can be done using state-space methods, starting with a realization of G and doing similarity transformation:

$$G(s) = [A, B, C, D]$$

$$= \left[\begin{bmatrix} A_1 & 0 \\ 0 & A_4 \end{bmatrix}, \begin{bmatrix} B_1 \\ B_2 \end{bmatrix}, [C_1 \ \ C_2], D \right]$$

(A_1 stable, A_4 antistable)

$$= D + [A_1, B_1, C_1, 0] + [A_4, B_2, C_2, 0] .$$

The condition $G^\sim = G$ implies that

$$G_1^\sim + G_2^\sim = G_1 + G_2 ,$$

i.e.

$$G_1^\sim - G_2 = G_1 - G_2^\sim .$$

The two sides of this equation are analytic in different half-planes and are strictly proper. By Liouville's theorem both sides equal zero. Thus $G_2 = G_1^\sim$, so from (2)

$$G = D + G_1 + G_1^\sim . \tag{3}$$

Bring in a minimal realization

$$G_1(s) = [A_1, B_1, C_1, 0] .$$

Then

$$G_1^{\sim}(s) = [-A_1^T, -C_1^T, B_1^T, 0]$$

and

$$G(s) = [A, B, C, D],$$

where

$$A := \begin{bmatrix} A_1 & 0 \\ 0 & -A_1^T \end{bmatrix} \tag{4}$$

$$B := \begin{bmatrix} B_1 \\ -C_1^T \end{bmatrix} \tag{5}$$

$$C := [C_1 \quad B_1^T] . \tag{6}$$

We are now set up to do a canonical factorization of G. From (4) we see that

$$\mathbf{X}_+(A) = \mathrm{Im} \begin{bmatrix} 0 \\ I \end{bmatrix} . \tag{7}$$

As in Section 1 bring in

$$A^{\times} = A - BD^{-1}C$$

$$= \begin{bmatrix} A_1 - B_1 D^{-1} C_1 & -B_1 D^{-1} B_1^T \\ C_1^T D^{-1} C_1 & -(A_1 - B_1 D^{-1} C_1)^T \end{bmatrix} . \tag{8}$$

Comparison of (8) and (2.1) shows that $A^{\times}$ is a Hamiltonian matrix. Note that $A^{\times}$ has no eigenvalues on the imaginary axis (this follows from the assumption $G^{-1} \in \mathbf{RH}_\infty$), $B_1 D^{-1} B_1^T$ is positive semi-definite, and

$$(A_1 - B_1 D^{-1} C_1, B_1 D^{-1} B_1^T)$$

is controllable (because (A_1, B_1) is). It follows from Theorem 2.2 and (7) that $\mathbf{X}_-(A^{\times})$ and $\mathbf{X}_+(A)$ are complementary. So by Theorem 1.1 G has a canonical factorization.

By virtue of Corollary 2.1 there exists a matrix X such that

$$\mathbf{X}_-(A^\times) = \operatorname{Im} \begin{bmatrix} I \\ X \end{bmatrix}.$$

Defining

$$T := \begin{bmatrix} I & 0 \\ X & I \end{bmatrix},$$

we get

$$T^{-1}AT = \begin{bmatrix} A_1 & 0 \\ ? & -A_1^T \end{bmatrix}$$

(question mark means irrelevant)

$$T^{-1}B = \begin{bmatrix} B_1 \\ -(C_1^T + XB_1) \end{bmatrix}$$

$$CT = [C_1 + B_1^T X \quad B_1^T].$$

By analogy with (1.8) and (1.9) we have $G = G_+ G_-$, where

$$G_+(s) := [-A_1^T, -(C_1^T + XB_1), B_1^T, D]$$

$$G_-(s) := [A_1, B_1, D^{-1}(C_1 + B_1^T X), I].$$

Notice that $G_+ = G_-^\sim D$, so that $G = G_-^\sim D G_-$. Finally, since D is positive definite (by (1c)) it has a square root $D^{1/2}$. Redefining G_- as

$$G_-(s) = [A_1, B_1, D^{-1/2}(C_1 + B_1^T X), D^{1/2}], \tag{9}$$

we get the desired spectral factorization $G = G_-^\sim G_-$. □

Exercise 1. Prove that if (A, B) is controllable and D is positive definite, then

$$(A - BD^{-1}C, BD^{-1}B^T)$$

is controllable.

A matrix G satisfying (1) also has a *co-spectral factorization*:

$$G = G_- G_-^{\sim}$$

$$G_-, G_-^{-1} \in \mathbf{RH}_\infty .$$

To get such a factorization, do a spectral factorization of $H := G^T$:

$$H = H_-^{\sim} H_-$$

$$H_-, H_-^{-1} \in \mathbf{RH}_\infty .$$

Then set $G_- = H_-^T$.

In applications one may want to do spectral factorization of a matrix given in special form. The method of proof in Theorem 1 can be used to solve the next exercises.

Exercise 2. Suppose $G \in \mathbf{RH}_\infty$ and $G(j\omega)$ has full column rank for all $0 \leq \omega \leq \infty$. Derive the following procedure for a spectral factorization of $G^{\sim} G$: obtain a minimal realization,

$$G(s) = [A, B, C, D];$$

define

$$\underline{D} := D^T D$$

$$H := \begin{bmatrix} A - B\underline{D}^{-1} D^T C & -B\underline{D}^{-1} B^T \\ -C^T C + C^T D\underline{D}^{-1} D^T C & -(A - B\underline{D}^{-1} D^T C)^T \end{bmatrix}$$

and find the unique matrix X such that

$$\mathbf{X}_-(H) = \mathrm{Im} \begin{bmatrix} I \\ X \end{bmatrix};$$

then

$$[A, B, \underline{D}^{-1/2}(D^T C + B^T X), \underline{D}^{1/2}]$$

is a spectral factor of $G^{\sim} G$.

Exercise 3. Suppose $G(s) = [A, B, C, I]$ is a minimal realization. Assume A is antistable and $A - BC$ is stable. Derive a formula for a spectral factor of

$G^{\sim}G$.

Exercise 4. Suppose $G \in \mathbf{RH}_\infty$ and $\gamma > \|G\|_\infty$. Derive a formula for a spectral factor of $\gamma^2 - G^{\sim}G$ starting from a minimal realization of G.

7.4 Inner-Outer Factorization

A matrix G in $\mathbf{RH}_\infty$ is *inner* if $G^{\sim}G = I$. This generalizes the definition of an inner function. Observe that for G to be inner it must be tall (number of rows $\geq$ number of columns). A useful property of an inner matrix is that its Laurent operator preserves inner products:

$$<Gf, Gh> = <f, h>, \quad f, h \in \mathbf{L}_2 .$$

It follows from this that pre-multiplication of an $\mathbf{L}_\infty$-matrix by G preserves norms:

$$F \in \mathbf{L}_\infty \Rightarrow \|GF\|_\infty = \|F\|_\infty .$$

An example of an inner matrix is

$$\begin{bmatrix} \dfrac{s+1}{s+\sqrt{2}} \\ \dfrac{1}{s+\sqrt{2}} \end{bmatrix} .$$

A matrix G in $\mathbf{RH}_\infty$ is *outer* if, for every Re $s > 0$, $G(s)$ has full row rank; equivalently, G has a right-inverse which is analytic in Re $s > 0$. This generalizes the definition of an outer function. An outer matrix is wide (number of rows $\leq$ number of columns). Of course, if G is square and

$$G, G^{-1} \in \mathbf{RH}_\infty ,$$

then G is outer. Another example is that $[F \quad G]$ is outer if F and G are left-coprime $\mathbf{RH}_\infty$-matrices.

An *inner-outer factorization* of a matrix G in $\mathbf{RH}_\infty$ is a factorization

$$G = G_i G_o$$

G_i inner, G_o outer.

It's a fact that every matrix in $\mathbf{RH}_\infty$ has an inner-outer factorization. We shall only need the following, slightly weaker result.

Theorem 1. If G is a matrix in $\mathbf{RH}_\infty$ and rank $G(j\omega)$ is constant for all $0 \leq \omega \leq \infty$, then G has an inner-outer factorization with the outer factor being right-invertible in $\mathbf{RH}_\infty$.

Proof. By Lemma 4.3.1 there exist square matrices G_1, H, K in $\mathbf{RH}_\infty$ satisfying the equation

$$G = H \begin{bmatrix} G_1 & 0 \\ 0 & 0 \end{bmatrix} K$$

and having the properties

$$H^{-1}, K^{-1} \in \mathbf{RH}_\infty$$

$$G_1 \text{ nonsingular.}$$

The rank assumption on G implies that $G_1(j\omega)$ is nonsingular for all $0 \leq \omega \leq \infty$, i.e. $G_1^{-1} \in \mathbf{RL}_\infty$. Define

$$F := H \begin{bmatrix} G_1 \\ 0 \end{bmatrix},$$

so that

$$G = [F \quad 0] K .$$

Now $F^\sim F$ has the properties

$$F^\sim F, (F^\sim F)^{-1} \in \mathbf{RL}_\infty$$

$$(F^\sim F)^\sim = F^\sim F$$

$$(F^\sim F)(\infty) > 0 ,$$

so it has a spectral factorization by Theorem 3.1:

$$F^\sim F = F_o^\sim F_o$$

$$F_o, F_o^{-1} \in \mathbf{RH}_\infty .$$

Define

$$G_i := FF_o^{-1}$$

$$G_o := [F_o \quad 0]K \ .$$

Then G_i is inner, G_o is right-invertible in $\mathbf{RH}_\infty$, and $G = G_i G_o$. □

Inner-outer factorization is relatively easy when $G(j\omega)$ has full column rank for all $0 \leq \omega \leq \infty$: let G_o be a spectral factor of $G^\sim G$ and then set $G_i := GG_o^{-1}$.

Example 1.

To do an inner-outer factorization of

$$G(s) = \begin{bmatrix} \frac{1}{s+1} \\ \frac{10s}{s+2} \\ \frac{s-1}{s+1} \end{bmatrix}$$

we follow Exercise 3.2 to get a spectral factor of $G^\sim G$. We have

$$G(s) = [A, B, C, D]$$

$$A = \begin{bmatrix} -3 & -2 \\ 1 & 0 \end{bmatrix}, \quad B = \begin{bmatrix} 1 \\ 0 \end{bmatrix}$$

$$C = \begin{bmatrix} 1 & 2 \\ -20 & -20 \\ -2 & -4 \end{bmatrix}, \quad D = \begin{bmatrix} 0 \\ 10 \\ 1 \end{bmatrix}.$$

Then

$$\underline{D} := D^T D = 101$$

$$H := \begin{bmatrix} A - B\underline{D}^{-1}D^T C & -B\underline{D}^{-1}B^T \\ -C^T C + C^T D\underline{D}^{-1}D^T C & -(A - B\underline{D}^{-1}D^T C)^T \end{bmatrix}$$

$$= \begin{bmatrix} -1 & .0198 & -.0099 & 0 \\ 1 & 0 & 0 & 0 \\ -1 & -2 & 1 & -1 \\ -2 & -7.9604 & -.0198 & 0 \end{bmatrix},$$

and

$$\mathbf{X}_-(H) = \text{Im} \begin{bmatrix} I \\ X \end{bmatrix}$$

where

$$X = \begin{bmatrix} 27.25 & 30.43 \\ 30.43 & 36.09 \end{bmatrix}.$$

Thus

$$G_o(s) = [A, B, \underline{D}^{-1/2}(D^T C + B^T X), \underline{D}^{1/2}]$$

$$= 10.05 \frac{(s+.9837)(s+.2861)}{(s+1)(s+2)}.$$

Finally

$$G_i(s) = G(s) G_o(s)^{-1}$$

$$= \frac{1}{10.05(s+.9837)(s+.2861)} \begin{bmatrix} s+2 \\ 10s(s+1) \\ (s-1)(s+2) \end{bmatrix}.$$

A matrix G is said to be *co-inner* or *co-outer* if G^T is inner or outer respectively. A *co-inner-outer factorization* has the form

$$G = G_{co} G_{ci}$$

G_{co} co–outer, G_{ci} co–inner.

An inner-outer factorization of G^T yields a co-inner-outer factorization of G.

7.5 J-Spectral Factorization

In this section we look at a rather special factorization required in Section 8.3. We start with a real-rational matrix $G_1(s)$ having the properties

$$G_1 \text{ is strictly proper,} \tag{1a}$$

$$G_1 \text{ is analytic in } \text{Re}\, s \leq 0, \tag{1b}$$

$$\|\Gamma_{G_1}\| < 1. \tag{1c}$$

Define the matrices

$$G := \begin{bmatrix} I & G_1 \\ 0 & I \end{bmatrix} \tag{2}$$

$$J := \begin{bmatrix} I & 0 \\ 0 & -I \end{bmatrix} \tag{3}$$

(not the same J as in Section 2). Our goal is to achieve the following *J-spectral factorization* of G:

$$G^{\sim} JG = G_{-}^{\sim} JG_{-} \tag{4}$$

$$G_{-}, G_{-}^{-1} \in \mathbf{RH}_{\infty} \; .$$

Bring in a minimal realization

$$G_1(s) = [A_1, B_1, C_1, 0] \; .$$

Note that A_1 is antistable. Defining

$$A := \begin{bmatrix} -A_1^T & C_1^T C_1 \\ 0 & A_1 \end{bmatrix}$$

$$B := \begin{bmatrix} C_1^T & 0 \\ 0 & B_1 \end{bmatrix}$$

$$C := \begin{bmatrix} 0 & C_1 \\ -B_1^T & 0 \end{bmatrix}$$

we have

$$(G^{\sim} JG)(s) = [A, B, C, J] \; .$$

Also

$$A^{\times} := A - BJ^{-1}C$$

$$= \begin{bmatrix} -A_1^T & 0 \\ -B_1 B_1^T & A_1 \end{bmatrix} .$$

Bring in the controllability and observability gramians:

$$A_1 L_c + L_c A_1^T = B_1 B_1^T$$

$$A_1^T L_o + L_o A_1 = C_1^T C_1 .$$

As in Theorem 2.1 we have

$$\mathbf{X}_-(A^\times) = \mathrm{Im} \begin{bmatrix} I \\ L_c \end{bmatrix}, \quad \mathbf{X}_+(A) = \mathrm{Im} \begin{bmatrix} L_o \\ I \end{bmatrix}.$$

Exercise 1. Show that assumption (1c) implies that $\mathbf{X}_-(A^\times)$ and $\mathbf{X}_+(A)$ are complementary.

From this exercise we can proceed with a canonical factorization of $G^\sim JG$. Defining

$$T := \begin{bmatrix} I & L_o \\ L_c & I \end{bmatrix}$$

$$N := (I - L_o L_c)^{-1} ,$$

we get (after some algebra)

$$T^{-1} A T = \begin{bmatrix} -N A_1^T N^{-1} & 0 \\ ? & A_1 \end{bmatrix}$$

$$T^{-1} B = \begin{bmatrix} N C_1^T & -N L_o B_1 \\ -L_c N C_1^T & N^T B_1 \end{bmatrix}$$

$$C T = \begin{bmatrix} C_1 L_c & C_1 \\ -B_1^T & -B_1^T L_o \end{bmatrix} .$$

Then we obtain from (1.8) and (1.9) that $G^\sim JF = G_+ G_-$, where

$$G_+(s) = [A_1, [-L_c N C_1^T \quad N^T B_1], \begin{bmatrix} C_1 \\ -B_1^T L_o \end{bmatrix}, J]$$

$$G_-(s) = [-N A_1^T N^{-1}, [N C_1^T \quad -N L_o B_1], \begin{bmatrix} C_1 L_c \\ B_1^T \end{bmatrix}, I] . \tag{5}$$

It can be checked that $G_+ = G_-^{\sim} J$, so that $G^{\sim} JG = G_-^{\sim} JG_-$. For future reference we record that

$$G_-(s)^{-1} = [-A_1^T, [NC_1^T \quad -NL_o B_1], \begin{bmatrix} -C_1 L_c \\ -B_1^T \end{bmatrix}, I], \tag{6}$$

which follows from (5) and some algebra.

Notes and References

This chapter is based primarily on Bart, Gohberg, and Kaashoek (1979), Doyle (1984), and Ball and Ran (1986): Theorem 1.1 is from Theorem 1.5 of Bart *et al.* (1979); Theorem 2.2 and Corollary 2.1 are based on Section 2.3.3 of Doyle (1984); Section 3 is based on Section 2.3.4 of Doyle (1984); and Section 5 is from Ball and Ran (1986).

Golub and Van Loan (1983) is an excellent source for numerical linear algebra, in particular the Schur decomposition. For the standard results on Lyapunov and Riccati equations see, for example, Wonham (1985). A standard reference for spectral factorization is Youla (1961). For a general treatment of inner-outer factorizations see Sz.-Nagy and Foias (1970).

CHAPTER 8

MODEL-MATCHING THEORY: PART II

This chapter treats the model-matching problem in the matrix-valued case.

8.1 Reduction to the Nehari Problem

The model-matching problem is a lot harder when the T_i's are matrix-valued functions than it is when they are scalar-valued. This section develops a high level algorithm for reducing the model-matching problem to the Nehari problem of approximating an $\mathbf{RL}_\infty$-matrix by an $\mathbf{RH}_\infty$-matrix; the Nehari problem will then be treated in Section 3.

Throughout this chapter the rank conditions of Theorem 6.1.1 are assumed to hold.

The problem is sufficiently hard that we shall content ourselves with accomplishing the following: to compute an upper bound γ for α such that $\gamma-\alpha$ is less than a pre-specified tolerance; and then to compute a Q in $\mathbf{RH}_\infty$ satisfying

$$||T_1-T_2QT_3||_\infty \leq \gamma . \tag{1}$$

Such a Q may not be optimal, but it will be as near optimality as we wish.

To see the development more clearly, let's first do the case $T_3=I$. Bring in an inner-outer factorization of T_2,

$$T_2 = U_i U_o$$

$$U_i \text{ inner,} \quad U_o \text{ outer ,}$$

and define the $\mathbf{RL}_\infty$-matrix

$$Y := (I-U_i U_i^\sim)T_1 .$$

If γ is a real number greater than $||Y||_\infty$, then the matrix $\gamma^2 - Y^\sim Y$ has a spectral factor Y_o. Define the $\mathbf{RL}_\infty$-matrix

$$R := U_i^{\sim} T_1 Y_o^{-1} .$$

Thus R depends on γ.

Theorem 1. $(T_3=I)$

(i) $\alpha = \inf\{\gamma : \|Y\|_\infty < \gamma,\ \text{dist}(R\,, \mathbf{RH}_\infty) < 1\}$

(ii) Suppose

$$\gamma > \alpha, \quad Q\,, X \in \mathbf{RH}_\infty$$

$$\|R - X\|_\infty \leq 1 \tag{2}$$

$$X = U_o\, Q Y_o^{-1} . \tag{3}$$

Then $\|T_1 - T_2 Q\|_\infty \leq \gamma$.

Part (i) of the theorem affords a method for computing an upper bound γ for α; then part (ii) yields a procedure for computing a nearly optimal Q. A particularly easy case of the theorem is when T_2 is square and nonsingular. Then U_i is square, $Y=0$, $Y_o = \gamma I$, $R = \gamma^{-1} U_i^{\sim} T_1$, and part (i) reduces to

$$\alpha = \text{dist}(U_i^{\sim} T_1, \mathbf{RH}_\infty) ,$$

i.e. α equals the norm of the Hankel operator with symbol $U_i^{\sim} T_1$. This is just like the scalar-valued case of Section 6.2.

The proof of Theorem 1 requires two preliminary technical facts.

Lemma 1. Let U be an inner matrix and define the $\mathbf{RL}_\infty$-matrix

$$E := \begin{bmatrix} U^{\sim} \\ I - UU^{\sim} \end{bmatrix} .$$

Then $\|EG\|_\infty = \|G\|_\infty$ for all matrices G in $\mathbf{RL}_\infty$.

Proof. It suffices to show that $E^{\sim} E = I$. But this follows easily from the fact that $U^{\sim} U = I$:

$$E^{\sim} E = [U \quad I - UU^{\sim}] \begin{bmatrix} U^{\sim} \\ I - UU^{\sim} \end{bmatrix}$$

$$= UU^{\sim} + (I - UU^{\sim})(I - UU^{\sim})$$

$$= I \ . \quad \square$$

Lemma 2. Suppose F and G are $\mathbf{RL}_{\infty}$-matrices with equal number of columns. If

$$\left\| \begin{bmatrix} F \\ G \end{bmatrix} \right\|_{\infty} < \gamma \, , \tag{4}$$

then

$$\|G\|_{\infty} < \gamma \tag{5}$$

and

$$\|FG_o^{-1}\|_{\infty} < 1 \, , \tag{6}$$

where G_o is a spectral factor of $\gamma^2 - G^{\sim}G$. Conversely, if (5) holds and

$$\|FG_o^{-1}\|_{\infty} \leq 1 \, ,$$

then

$$\left\| \begin{bmatrix} F \\ G \end{bmatrix} \right\|_{\infty} \leq \gamma \, .$$

Proof. We'll prove the first statement. Assume (4). Then (5) follows immediately because

$$\|G\|_{\infty} \leq \left\| \begin{bmatrix} F \\ G \end{bmatrix} \right\|_{\infty} \, .$$

It follows in turn from Theorem 7.3.1 that $\gamma^2 - G^{\sim}G$ has a spectral factorization:

$$\gamma^2 - G^{\sim}G = G_o^{\sim}G_o \tag{7}$$

$$G_o \, , G_o^{-1} \in \mathbf{RH}_{\infty} \, .$$

To prove (6), define

$$\epsilon := \gamma - \left\| \begin{bmatrix} F \\ G \end{bmatrix} \right\|_\infty , \tag{8}$$

let f be an $\mathbf{L}_2$-vector of unit norm, and define $g := G_o^{-1} f$. Starting from (8) we have in succession

$$\left\| \begin{bmatrix} F \\ G \end{bmatrix} g \right\|_2 \leq (\gamma - \epsilon) \| g \|_2$$

$$\left\langle \begin{bmatrix} F \\ G \end{bmatrix} g , \begin{bmatrix} F \\ G \end{bmatrix} g \right\rangle \leq (\gamma - \epsilon)^2 < g , g >$$

$$< g , (F^{\sim} F + G^{\sim} G) g > \leq \gamma^2 < g , g > - \epsilon (2\gamma - \epsilon) \| g \|_2^2$$

$$< g , F^{\sim} F g > \leq < g , (\gamma^2 - G^{\sim} G) g > - \epsilon (2\gamma - \epsilon) \| G_o \|_\infty^{-2} .$$

The last step used the inequality

$$1 = \| G_o \, g \|_2 \leq \| G_o \|_\infty \| g \|_2 .$$

Now using (7) we get

$$\| F g \|_2^2 \leq \| G_o \, g \|_2^2 - \epsilon (2\gamma - \epsilon) \| G_o \|_\infty^{-2} .$$

Hence

$$\| F G_o^{-1} f \|_2^2 \leq 1 - \epsilon (2\gamma - \epsilon) \| G_o \|_\infty^{-2} .$$

Since f was arbitrary we find that

$$\| F G_o^{-1} \|_\infty^2 \leq 1 - \epsilon (2\gamma - \epsilon) \| G_o \|_\infty^{-2} .$$

Since $\epsilon(2\gamma - \epsilon) > 0$, we arrive at (6). □

Exercise 1. Prove the converse in Lemma 2.

Proof of Theorem 1.

(i) Let

$$\gamma_{inf} := \inf \{ \gamma : \| Y \|_\infty < \gamma, \, \mathrm{dist}(R , \mathbf{RH}_\infty) < 1 \} .$$

Choose $\epsilon > 0$ and then choose γ such that $\alpha + \epsilon > \gamma > \alpha$. Then there exists Q in $\mathbf{RH}_\infty$ such that

$$\| T_1 - T_2 Q \|_\infty < \gamma .$$

Equivalently, from Lemma 1

$$\left\| \begin{bmatrix} U_i^{\sim} \\ I - U_i U_i^{\sim} \end{bmatrix} (T_1 - T_2 Q) \right\|_\infty < \gamma . \tag{9}$$

Now

$$\begin{bmatrix} U_i^{\sim} \\ I - U_i U_i^{\sim} \end{bmatrix} T_2 = \begin{bmatrix} U_o \\ 0 \end{bmatrix},$$

so (9) is equivalent to

$$\left\| \begin{bmatrix} U_i^{\sim} T_1 - U_o Q \\ Y \end{bmatrix} \right\|_\infty < \gamma .$$

This implies from Lemma 2 that

$$\| Y \|_\infty < \gamma \tag{10}$$

$$\| U_i^{\sim} T_1 Y_o^{-1} - U_o Q Y_o^{-1} \|_\infty < 1 . \tag{11}$$

The latter inequality implies

$$\text{dist}(R \, , \, U_o \, \mathbf{RH}_\infty Y_o^{-1}) < 1 . \tag{12}$$

But U_o is right-invertible in $\mathbf{RH}_\infty$ (Theorem 7.4.1) and Y_o is invertible in $\mathbf{RH}_\infty$. Therefore

$$U_o \, \mathbf{RH}_\infty Y_o^{-1} = \mathbf{RH}_\infty ,$$

so (12) gives

$$\text{dist}(R \, , \, \mathbf{RH}_\infty) < 1 . \tag{13}$$

From (10), (13), and the definition of γ_{inf} we conclude that $\gamma_{inf} \leq \gamma$. Thus $\gamma_{inf} < \alpha + \epsilon$. Since ϵ was arbitrary, $\gamma_{inf} \leq \alpha$.

For the reverse inequality, again choose $\epsilon > 0$ and then choose γ such that $\gamma_{inf} + \epsilon > \gamma > \gamma_{inf}$. Then (10) and (13) hold, so (11) holds for some Q in $\mathbf{RH}_\infty$. Lemma 2 now implies that

$$\left\| \begin{bmatrix} U_i^{\sim} T_1 - U_o Q \\ Y \end{bmatrix} \right\|_\infty \leq \gamma .$$

Finally, this leads (as above) to

$$||T_1 - T_2 Q||_\infty \leq \gamma .$$

Thus $\alpha \leq \gamma < \gamma_{inf} + \epsilon$, so $\alpha \leq \gamma_{inf}$.

(ii) This part follows from the previous paragraph. □

We saw in Section 6.2 that

$$\text{dist}(R\,,\,\mathbf{RH}_\infty) = \text{dist}(R\,,\,\mathbf{H}_\infty)$$

for a scalar-valued R in $\mathbf{RL}_\infty$. This is true in the matrix-valued case too.

Lemma 3. For R in $\mathbf{RL}_\infty$

$$\text{dist}(R\,,\,\mathbf{RH}_\infty) = \text{dist}(R\,,\,\mathbf{H}_\infty) = ||\Gamma_R|| .$$

Proof. We have

$$\text{dist}(R\,,\,\mathbf{RH}_\infty) \geq \text{dist}(R\,,\,\mathbf{H}_\infty) = ||\Gamma_R|| ,$$

the latter equality being Nehari's theorem. Choose $\epsilon > 0$ and set $\beta := ||\Gamma_R||$. Then

$$\begin{aligned} \text{dist}[(\beta+\epsilon)^{-1}R\,,\,\mathbf{H}_\infty] &= (\beta+\epsilon)^{-1}||\Gamma_R|| \\ &= \beta/(\beta+\epsilon) \\ &< 1 . \end{aligned}$$

As we shall see in Section 3, this inequality implies there exists X in $\mathbf{RH}_\infty$ such that

$$||(\beta+\epsilon)^{-1}R - X||_\infty \leq 1 .$$

Thus

$$\begin{aligned} \text{dist}(R\,,\,\mathbf{RH}_\infty) &\leq \beta+\epsilon \\ &= \text{dist}(R\,,\,\mathbf{H}_\infty)+\epsilon . \end{aligned}$$

Since ϵ was arbitrary

$$\text{dist}(R\,,\,\mathbf{RH}_\infty) \leq \text{dist}(R\,,\,\mathbf{H}_\infty) . \quad □$$

It can also be proved that the distance from R to $\mathbf{RH}_\infty$ is achieved, i.e. there exists X in $\mathbf{RH}_\infty$ such that

$$||R - X||_\infty = \text{dist}(R\,,\,\mathbf{RH}_\infty)\,.$$

Based on Theorem 1 and Lemma 3 we have the following high level algorithm for finding nearly optimal Q's in the case $T_3 = I$.

Step 1. Compute Y and $||Y||_\infty$.

Step 2. Find an upper bound α_1 for α.

Step 3. Select a trial value for γ in the interval $(||Y||_\infty, \alpha_1]$.

Step 4. Compute R and $||\Gamma_R||$. Then $||\Gamma_R|| < 1$ iff $\alpha < \gamma$, so increase or decrease the value of γ accordingly and return to Step 3. When a sufficiently accurate upper bound for α is obtained, continue to Step 5.

Step 5. Find a matrix X in $\mathbf{RH}_\infty$ such that $||R - X||_\infty \le 1$.

Step 6. Solve $X = U_o\, Q Y_o^{-1}$ for Q in $\mathbf{RH}_\infty$.

Step 1 involves implementation of the procedures in the previous chapter for spectral factorization and inner-outer factorization. For Step 2 the simplest bound would be $\alpha_1 = ||T_1||_\infty$. Then binary search could be used to iterate on γ. Step 6 is not too difficult if U_o is square; this happens when T_2 is tall, which frequently can be arranged by suitable problem formulation. When U_o isn't square, Step 6 would be more difficult. How to do Step 5 is the subject of the next two sections.

Example 1.

Let's continue with the tracking example of Chapter 3, again taking

$$P(s) = \frac{s-1}{s\,(s-2)}$$

$$W(s) = \frac{s+1}{10s+1}$$

and $\rho=1$. In Section 4.5 we obtained

$$T_1(s) = T_3(s) = \begin{bmatrix} \frac{s+1}{10s+1} \\ 0 \end{bmatrix}$$

$$T_2(s) = \begin{bmatrix} -\frac{s-1}{s^2+s+1} \\ \frac{s(s-2)}{s^2+s+1} \end{bmatrix}.$$

The matrix Q is of dimension 1×2:

$$Q = [Q_1 \quad Q_2].$$

Notice that $QT_3 = WQ_1$. Thus by the substitution

$$T_2 \leftarrow WT_2$$

we arrive at a model-matching problem with $T_3 = I$, namely

$$\text{minimize } \|T_1 - T_2 Q_1\|_\infty, \quad Q_1 \in \mathbf{RH}_\infty.$$

The previous algorithm produces the following results.

Step 1.

$$U_i(s) = \frac{1}{s^2+\sqrt{7}s+1} \begin{bmatrix} -s+1 \\ s(s-2) \end{bmatrix}$$

$$U_o(s) = \frac{(s^2+\sqrt{7}s+1)(s+1)}{(s^2+s+1)(10s+1)}$$

$$Y(s) = \frac{s+1}{(10s+1)(s^4-5s^2+1)} \begin{bmatrix} s^2(s^2-4) \\ -s(s+1)(s-2) \end{bmatrix}$$

$$\|Y\|_\infty = 0.1683$$

Step 2.

$$\alpha_1 = \|T_1\|_\infty = 1$$

Steps 3 and 4. From the first two steps we know that α lies in the interval $[.1683, 1]$. We now use binary search on γ in this interval, testing if $||\Gamma_R|| < 1$. To locate α within, say, 3% of the length of this interval, we need five iterations:

γ	$\|\|\Gamma_R\|\|$
0.5846	0.3710
0.3774	0.5996
0.2729	0.8952
0.2206	1.2327
0.2468	1.0317

We conclude that α lies in the interval $[.2468, .2729]$. Let's proceed with γ=.2729. Then a spectral factor of $\gamma^2 - Y^{\sim} Y$ (calculated via Theorem 7.3.1) is

$$Y_o(s) = .2539 \frac{s^3 + 2.678s^2 + 1.081s + .1075}{(s+.1)(s^2+\sqrt{7}s+1)} .$$

There follows

$$R(s) = \frac{.3938(s+1)^2(s^2+\sqrt{7}s+1)}{(s^2-\sqrt{7}s+1)(s^3+2.678s^2+1.081s+.1075)} .$$

(The values of $||\Gamma_R||$ in the above table are calculated as in Example 5.2.1.)

Step 5. Since R is scalar-valued, Theorem 6.2.1 can be applied to find the closest function X in $\mathbf{RH}_\infty$.

$$R(s) = [A, B, C, D] + (\text{a function in } \mathbf{RH}_\infty)$$

$$A = \begin{bmatrix} 2.189 & 0 \\ 0 & .4569 \end{bmatrix}$$

$$B = \begin{bmatrix} 1.038 \\ -.9291 \end{bmatrix}$$

$$C = [1 \quad 1]$$

$$L_c = \begin{bmatrix} .2463 & -.3646 \\ -.3646 & .9448 \end{bmatrix}$$

$$L_o = \begin{bmatrix} .2284 & .3780 \\ .3780 & 1.095 \end{bmatrix}$$

$$\lambda := \text{maximum eigenvalue of } L_c L_o = .8952$$

$$w := \text{corresponding eigenvector} = \begin{bmatrix} -.3466 \\ 1 \end{bmatrix}$$

$$v := \lambda^{-1} L_o w = \begin{bmatrix} .3338 \\ 1.076 \end{bmatrix}$$

$$X(s) = R(s) - \lambda[A, w, C, 0] \times [-A^T, v, B^T, 0]^{-1}$$

$$= .8952 \frac{(s^2+\sqrt{7}s+1)(s^2+2.656s+1.033)}{(s+3.108)(s^3+2.678s^2+1.081s+.1075)}$$

Step 6.

$$Q = [Q_1 \; Q_2]$$

$$Q_1(s) = X(s) Y_o(s) U_o(s)^{-1}$$

$$= 2.273 \frac{(s^2+s+1)(s^2+2.656s+1.033)}{(s+1)(s+3.108)(s^2+\sqrt{7}s+1)}$$

The function Q_2 is unconstrained, and hence may be set to zero.

Finally, the controller K is computed using formula (4.5.2).

$$K = [C_1 \; C_2]$$

$$C_1(s) = -2.273 \frac{(s+1)(s^2+s+1)(s^2+2.656s+1.033)}{(s+3.108)(s^2+\sqrt{7}s+1)(s^2+5s-18)}$$

$$C_2(s) = -\frac{32s-1}{s^2+5s-18}$$

Notice that C_1 is unstable, so the controller can't be implemented as shown in Figure 4 of Chapter 3. (This is a case where every stabilizing controller is itself unstable.) The theory guarantees, however, that C_2 contains the unstable factor of C_1. This unstable factor would be moved past the summing junction into the loop.

The properties of this design are illustrated in the Bode magnitude plot of Figure 1. The transfer function, say H_1, from reference r to tracking error $r-v$ has magnitude less than -10 db over the frequency range [0, .1], approximately the bandwidth of the weighting function W (smaller tracking error could be

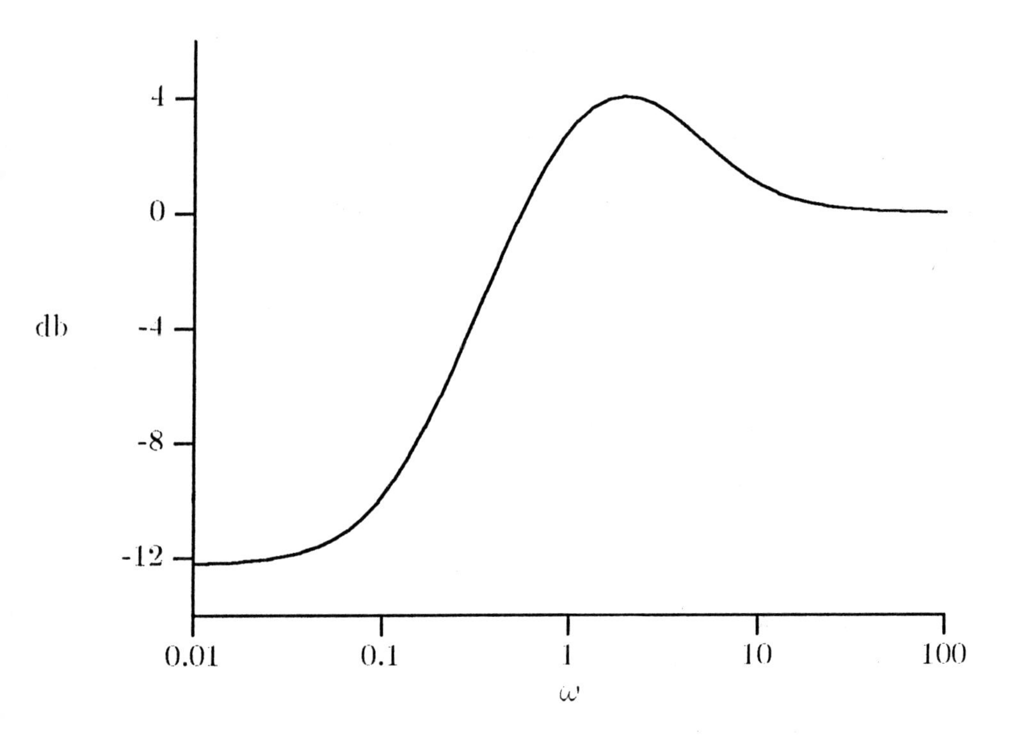

Figure 8.1.1. Magnitude Bode plot of H_1, Example 1

obtained by reducing the weighting ρ on control energy), it peaks to about 4 db, and it rolls off to 0 db at high frequency, as it must for a proper controller. The actual quantity being minimized is the $\mathbf{H}_\infty$-norm of the transfer matrix

$$\begin{bmatrix} H_1 W \\ H_2 W \end{bmatrix},$$

where H_2 is the transfer function from r to u. This norm equals the supremum of

$$(|H_1(j\omega)|^2 + |H_2(j\omega)|^2)^{1/2} |W(j\omega)|$$

over all ω. For our design this function is very nearly flat at -11.3 db.

The generalization when $T_3 \neq I$ uses the following definitions:

$$T_2 = U_i U_o, \quad U_i \text{ inner}, \quad U_o \text{ outer}$$

$$Y := (I - U_i U_i^\sim) T_1$$

$$Y_o = \text{spectral factor of } \gamma^2 - Y^\sim Y$$

$$T_3 Y_o^{-1} = V_{co} V_{ci}, \quad V_{co} \text{ co-outer}, \quad V_{ci} \text{ co-inner}$$

$$Z := U_i^\sim T_1 Y_o^{-1} (I - V_{ci}^\sim V_{ci})$$

$$Z_{co} = \text{co-spectral factor of } I - ZZ^\sim$$

$$R := Z_{co}^{-1} U_i^\sim T_1 Y_o^{-1} V_{ci}^\sim .$$

Notice that $R, Y, Z \in \mathbf{RL}_\infty$ and $X \in \mathbf{RH}_\infty$; Y is a function of T_1 and T_2; R and Z are functions of γ, T_1, T_2, and T_3. The matrix $Z_{co}^{-1} U_o$ is right-invertible over $\mathbf{RH}_\infty$ and V_{co} is left-invertible over $\mathbf{RH}_\infty$.

Theorem 2.

(i) $\alpha = \inf\{\gamma : \|Y\|_\infty < \gamma, \|Z\|_\infty < 1, \text{dist}(R, \mathbf{RH}_\infty) < 1\}$.

(ii) Suppose

$$\gamma > \alpha, \quad Q, X \in \mathbf{RH}_\infty$$

$$\|R - X\|_\infty \leq 1$$

$$X = Z_{co}^{-1} U_o \, Q V_{co} \; .$$

Then $||T_1 - T_2 Q T_3||_\infty \leq \gamma$.

The proof is analogous to that of Theorem 1 and is therefore omitted. The general high level algorithm is as follows:

Step 1. Compute Y and $||Y||_\infty$.

Step 2. Find an upper bound α_1 for α.

Step 3. Select a trial value for γ in the interval $(||Y||_\infty, \alpha_1]$.

Step 4. Compute Z and $||Z||_\infty$.

Step 5. If $||Z||_\infty < 1$, continue; if not, increase γ and return to Step 4.

Step 6. Compute R and $||\Gamma_R||$. Then $||\Gamma_R|| < 1$ iff $\alpha < \gamma$, so increase or decrease the value of γ accordingly and return to Step 3. When a sufficiently accurate upper bound for α is obtained, continue.

Step 7. Find a matrix X in $\mathbf{RH}_\infty$ such that $||R - X||_\infty \leq 1$.

Step 8. Solve $X = Z_{co}^{-1} U_o \, Q V_{co}$ for Q in $\mathbf{RH}_\infty$.

8.2 Krein Space

This section introduces a geometric structure which will be used to solve the Nehari problem.

Let $\mathbf{X}$ and $\mathbf{Y}$ be two Hilbert spaces. There is a natural way to add them together to get a third Hilbert space, their *external direct sum* $\mathbf{X} \oplus \mathbf{Y}$. We shall represent vectors in $\mathbf{X} \oplus \mathbf{Y}$ like this: $\begin{pmatrix} x \\ y \end{pmatrix}$. As a set, $\mathbf{X} \oplus \mathbf{Y}$ consists of all such vectors as x ranges over $\mathbf{X}$ and y over $\mathbf{Y}$. Vector addition and scalar multiplication are defined componentwise, and the inner product is defined as follows:

$$\left\langle \begin{pmatrix} x_1 \\ y_1 \end{pmatrix}, \begin{pmatrix} x_2 \\ y_2 \end{pmatrix} \right\rangle := <x_1,x_2> + <y_1,y_2> .$$

Now introduce in addition an *indefinite inner-product* on $\mathbf{X} \oplus \mathbf{Y}$:

$$\left[\begin{pmatrix} x_1 \\ y_1 \end{pmatrix}, \begin{pmatrix} x_2 \\ y_2 \end{pmatrix} \right] := <x_1,x_2> - <y_1,y_2> .$$

This is indefinite because $[z,z]$ can be negative, zero, or positive, depending on the particular z in $\mathbf{X} \oplus \mathbf{Y}$. A more compact way of defining $[\ ,\ ]$ is to introduce the operator J on $\mathbf{X} \oplus \mathbf{Y}$:

$$J\begin{pmatrix} x \\ y \end{pmatrix} := \begin{pmatrix} x \\ -y \end{pmatrix} .$$

Then $[z_1,z_2] := <z_1,Jz_2>$. The external direct sum $\mathbf{X} \oplus \mathbf{Y}$ together with the above indefinite inner-product is called a *Krein space.*

A vector z in $\mathbf{X} \oplus \mathbf{Y}$ is *negative* if $[z,z] \leq 0$, and a subspace of $\mathbf{X} \oplus \mathbf{Y}$ is *negative* if all its vectors are negative.

Consider an operator Φ from $\mathbf{Y}$ to $\mathbf{X}$. Its *graph* is a subspace of $\mathbf{X} \oplus \mathbf{Y}$, namely,

$$\left\{ \begin{pmatrix} \Phi y \\ y \end{pmatrix} : y \in \mathbf{Y} \right\} .$$

It is an elementary fact (and is easy to prove) that the graph is a closed subset of $\mathbf{X} \oplus \mathbf{Y}$.

Example 1.

Let F be a matrix in $\mathbf{RH}_\infty$ and consider the compression to $\mathbf{H}_2$ of the Laurent operator with symbol F (i.e. the Toeplitz operator). Its graph is

$$\left\{ \begin{pmatrix} Fg \\ g \end{pmatrix} : g \in \mathbf{H}_2 \right\} ,$$

or equivalently

$$\begin{bmatrix} F \\ I \end{bmatrix} \mathbf{H}_2 ,$$

where we use the notation

$$M\mathbf{H}_2 := \{Mg \ : g \in \mathbf{H}_2\} .$$

This graph is a subspace of $\mathbf{H}_2 \oplus \mathbf{H}_2$. It's convenient to denote it by $\mathbf{G}_F$.

Example 2.

If $F \in \mathbf{RL}_\infty$, then the corresponding graph

$$\mathbf{G}_F := \begin{bmatrix} F \\ I \end{bmatrix} \mathbf{H}_2$$

lives in $\mathbf{L}_2 \oplus \mathbf{H}_2$.

8.3 The Nehari Problem

This section solves the Nehari problem posed as follows: given R in $\mathbf{RL}_\infty$ with $\text{dist}(R , \mathbf{RH}_\infty)<1$, find all X's in $\mathbf{RH}_\infty$ such that $||R - X||_\infty \leq 1$. Only some of these X's are closest to R, i.e. satisfy

$$||R - X||_\infty = \text{dist}(R , \mathbf{RH}_\infty) .$$

We saw in Lemma 1.3 that the distance equals $||\Gamma_R||$, so the standing assumption in this section is that $||\Gamma_R||<1$.

We may as well assume in addition that R is strictly proper and analytic in Re $s \leq 0$, i.e. $R^\sim \in \mathbf{RH}_\infty$. Otherwise, factor R uniquely as

$$R = R_1 + R_2$$

$$R_1^\sim , R_2 \in \mathbf{RH}_\infty , \quad R_1 \text{ strictly proper} .$$

We've already observed in Chapter 5 that $\Gamma_R = \Gamma_{R_1}$. So to solve the Nehari problem for R, solve it for R_1, i.e. find all X_1's in $\mathbf{RH}_\infty$ such that $||R_1 - X_1||_\infty \leq 1$, and then set $X = X_1 + R_2$.

We need a preliminary fact.

Lemma 1. Let $F \in \mathbf{RL}_\infty$. Then $F\,\mathbf{H}_2 \subset \mathbf{H}_2$ iff $F \in \mathbf{RH}_\infty$. If F is square and $F\,\mathbf{H}_2 = \mathbf{H}_2$, then $F^{-1} \in \mathbf{RH}_\infty$.

Proof. The implication

$$F \in \mathbf{RH}_\infty \Rightarrow F\,\mathbf{H}_2 \subset \mathbf{H}_2$$

is easy (and has already been noted in Theorem 2.4.2).

Suppose $F\,\mathbf{H}_2 \subset \mathbf{H}_2$. Since each column of the matrix $(s+1)^{-1}I$ belongs to $\mathbf{H}_2$, the same is true of $(s+1)^{-1}F(s)$. Therefore, this latter matrix is strictly proper and analytic in Re $s \geq 0$. Hence F is proper and analytic in Re $s \geq 0$, i.e. $F \in \mathbf{RH}_\infty$.

Finally, suppose F is square and $F\,\mathbf{H}_2 = \mathbf{H}_2$. Then there exists a matrix G, each of whose columns belongs to $\mathbf{H}_2$, such that

$$F(s)G(s) = (s+1)^{-1}I \ .$$

This implies that F has an inverse in $\mathbf{RH}_\infty$, namely, $(s+1)G(s)$. □

In terms of $S := R - X$, a problem equivalent to the Nehari problem is this: find all S's in $\mathbf{RL}_\infty$ such that $\|S\|_\infty \leq 1$ and $R - S \in \mathbf{RH}_\infty$. The next lemma gives geometric characterizations of these two conditions. Define the $\mathbf{RL}_\infty$-matrix

$$G := \begin{bmatrix} I & R \\ 0 & I \end{bmatrix} . \tag{1}$$

Lemma 2. Let $S \in \mathbf{RL}_\infty$. Then $\|S\|_\infty \leq 1$ iff $\mathbf{G}_S$ is negative, and $R - S \in \mathbf{RH}_\infty$ iff

$$\mathbf{G}_S \subset G(\mathbf{H}_2 \oplus \mathbf{H}_2) \ .$$

Proof. Suppose $\|S\|_\infty \leq 1$. A vector in $\mathbf{G}_S$ has the form $\begin{pmatrix} Sf \\ f \end{pmatrix}$ for some f in $\mathbf{H}_2$. This vector is negative:

$$\left[\begin{pmatrix} Sf \\ f \end{pmatrix}, \begin{pmatrix} Sf \\ f \end{pmatrix} \right] = ||Sf\,||_2^2 - ||f\,||_2^2$$

$$\leq (||S\,||_\infty^2 - 1)||f\,||_2^2$$

$$\leq 0 \; .$$

The converse is equally easy.

Now suppose $R - S \in \mathbf{RH}_\infty$. Then by Lemma 1

$$(S - R\,)\mathbf{H}_2 \subset \mathbf{H}_2 \; ,$$

so

$$\mathbf{G}_S = \begin{bmatrix} S \\ I \end{bmatrix} \mathbf{H}_2$$

$$= \begin{bmatrix} I & R \\ 0 & I \end{bmatrix} \begin{bmatrix} S - R \\ I \end{bmatrix} \mathbf{H}_2$$

$$\subset \begin{bmatrix} I & R \\ 0 & I \end{bmatrix} (\mathbf{H}_2 \oplus \mathbf{H}_2)$$

$$= G\,(\mathbf{H}_2 \oplus \mathbf{H}_2) \; .$$

Conversely, if

$$\mathbf{G}_S \subset G\,(\mathbf{H}_2 \oplus \mathbf{H}_2) \; ,$$

then

$$\begin{bmatrix} S \\ I \end{bmatrix} \mathbf{H}_2 \subset \begin{bmatrix} I & R \\ 0 & I \end{bmatrix} (\mathbf{H}_2 \oplus \mathbf{H}_2) \; .$$

Pre-multiply by $[-I \quad R\,]$ to get

$$(R - S\,)\mathbf{H}_2 \subset \mathbf{H}_2 \; ,$$

which implies by Lemma 1 that $R - S \in \mathbf{RH}_\infty$. □

In view of Lemma 2 we would like to be able to characterize negative graphs contained in $G\,(\mathbf{H}_2 \oplus \mathbf{H}_2)$. The J-spectral factorization of Section 7.5 was introduced for this very purpose. Following Section 7.5 we have

$$G^{\sim}JG = G_{-}^{\sim}JG_{-} \tag{2}$$

$$G_{-}, G_{-}^{-1} \in \mathbf{RH}_{\infty}$$

$$J := \begin{bmatrix} I & 0 \\ 0 & -I \end{bmatrix}.$$

Define

$$L := GG_{-}^{-1}. \tag{3}$$

Then from (2)

$$L^{\sim}JL = J \tag{4}$$

and from (3)

$$L(\mathbf{H}_2 \oplus \mathbf{H}_2) = G(\mathbf{H}_2 \oplus \mathbf{H}_2). \tag{5}$$

A square matrix M in $\mathbf{RL}_{\infty}$ having the property $M^{\sim}JM = J$ is said to be *J-unitary*. Such a matrix is invertible in $\mathbf{RL}_{\infty}$; in fact the inverse of M is $JM^{\sim}J$. The usefulness of J-unitary matrices derives from the fact that (under mild conditions) they map negative graphs into negative graphs. The precise statement is as follows.

Lemma 3. Let X be an $\mathbf{RL}_{\infty}$-matrix with $\|X\|_{\infty} \leq 1$. Suppose M is a J-unitary matrix having the properties

$$M\mathbf{G}_X \subset \mathbf{L}_2 \oplus \mathbf{H}_2 \tag{6}$$

$$\{0\} \oplus \mathbf{H}_2 \subset M(\mathbf{L}_2 \oplus \mathbf{H}_2). \tag{7}$$

Then there exists Y in $\mathbf{RL}_{\infty}$ such that $\|Y\|_{\infty} \leq 1$ and $\mathbf{G}_Y = M\mathbf{G}_X$.

Proof. Define

$$\begin{bmatrix} Y_1 \\ Y_2 \end{bmatrix} := M \begin{bmatrix} X \\ I \end{bmatrix}. \tag{8}$$

Then from (6)

$$\begin{bmatrix} Y_1 \\ Y_2 \end{bmatrix} \mathbf{H}_2 = M\mathbf{G}_X \subset \mathbf{L}_2 \oplus \mathbf{H}_2,$$

so pre-multiplying by $[0 \ I]$ we get $Y_2\mathbf{H}_2 \subset \mathbf{H}_2$. Then by Lemma 1, $Y_2 \in \mathbf{RH}_\infty$. We shall show that $Y_2^{-1} \in \mathbf{RH}_\infty$. This takes three steps.

Claim #1: $M\mathbf{G}_X$ is a negative subspace of $\mathbf{L}_2 \oplus \mathbf{H}_2$.

A vector in $M\mathbf{G}_X$ has the form

$$M\begin{pmatrix} Xf \\ f \end{pmatrix}$$

for some f in $\mathbf{H}_2$. Then

$$\left[M\begin{pmatrix} Xf \\ f \end{pmatrix}, M\begin{pmatrix} Xf \\ f \end{pmatrix}\right] = \left\langle M\begin{pmatrix} Xf \\ f \end{pmatrix}, JM\begin{pmatrix} Xf \\ f \end{pmatrix}\right\rangle$$

$$= \left\langle \begin{pmatrix} Xf \\ f \end{pmatrix}, M^{\sim}JM\begin{pmatrix} Xf \\ f \end{pmatrix}\right\rangle$$

$$= \left\langle \begin{pmatrix} Xf \\ f \end{pmatrix}, J\begin{pmatrix} Xf \\ f \end{pmatrix}\right\rangle$$

$$= \left[\begin{pmatrix} Xf \\ f \end{pmatrix}, \begin{pmatrix} Xf \\ f \end{pmatrix}\right].$$

The last quantity is ≤ 0 because $\mathbf{G}_X$ is negative.

Claim #2: $Y_2\mathbf{H}_2$ is a closed subspace of $\mathbf{H}_2$.

Suppose $\{f_k\}$ is a sequence in $Y_2\mathbf{H}_2$ which converges to some f in $\mathbf{H}_2$. Then

$$\begin{pmatrix} h_k \\ f_k \end{pmatrix} \in \begin{bmatrix} Y_1 \\ Y_2 \end{bmatrix}\mathbf{H}_2 = M\mathbf{G}_X$$

for certain vectors h_k in $\mathbf{L}_2$. Since $\{f_k\}$ is Cauchy and $M\mathbf{G}_X$ is negative, it follows that $\{h_k\}$ is Cauchy too, so it converges to some h in $\mathbf{L}_2$. Since $\mathbf{G}_X$ is closed and $M, M^{-1} \in \mathbf{RL}_\infty$, it follows that $M\mathbf{G}_X$ is closed. Thus

$$\begin{pmatrix} h \\ f \end{pmatrix} \in \begin{bmatrix} Y_1 \\ Y_2 \end{bmatrix}\mathbf{H}_2,$$

so $f \in Y_2\mathbf{H}_2$.

Claim #3: $Y_2\mathbf{H}_2 = \mathbf{H}_2$.

Suppose otherwise. Since $Y_2\mathbf{H}_2$ is closed, we have

$$\mathbf{H}_2 = (Y_2\mathbf{H}_2)^{\perp} \oplus (Y_2\mathbf{H}_2)\,.$$

Let g be a nonzero vector in $(Y_2\mathbf{H}_2)^{\perp}$ and define

$$\begin{pmatrix} f_1 \\ f_2 \end{pmatrix} := M^{-1}\begin{pmatrix} 0 \\ g \end{pmatrix}. \tag{9}$$

Then (7) implies that $f_2 \in \mathbf{H}_2$, so g and $Y_2 f_2$ are orthogonal. Now we get

$$\begin{aligned}
0 &= -\langle g, Y_2 f_2 \rangle \\
&= \left\langle \begin{pmatrix} 0 \\ g \end{pmatrix}, J\begin{pmatrix} Y_1 f_2 \\ Y_2 f_2 \end{pmatrix} \right\rangle \\
&= \left\langle M\begin{pmatrix} f_1 \\ f_2 \end{pmatrix}, JM\begin{pmatrix} Xf_2 \\ f_2 \end{pmatrix} \right\rangle \quad \text{from (8) and (9)} \\
&= \left\langle \begin{pmatrix} f_1 \\ f_2 \end{pmatrix}, M^{\sim}JM\begin{pmatrix} Xf_2 \\ f_2 \end{pmatrix} \right\rangle \\
&= \left\langle \begin{pmatrix} f_1 \\ f_2 \end{pmatrix}, J\begin{pmatrix} Xf_2 \\ f_2 \end{pmatrix} \right\rangle \\
&= \langle f_1, Xf_2 \rangle - \|f_2\|_2^2\,.
\end{aligned}$$

Thus

$$\begin{aligned}
\|f_2\|_2^2 &= \langle f_1, Xf_2 \rangle \\
&\le \|f_1\|_2 \|Xf_2\|_2 \\
&\le \|f_1\|_2 \|f_2\|_2\,,
\end{aligned}$$

or

$$\|f_2\|_2 \le \|f_1\|_2\,. \tag{10}$$

But $\begin{pmatrix} 0 \\ g \end{pmatrix}$ is strictly negative and M^{-1} is J-unitary. This implies that $\begin{pmatrix} f_1 \\ f_2 \end{pmatrix}$ is strictly negative too, i.e.

$$\|f_1\|_2 < \|f_2\|_2\,. \tag{11}$$

Inequalities (10) and (11) are contradictory.

It follows from the third claim and Lemma 1 that $Y_2^{-1} \in \mathbf{RH}_\infty$. Defining $Y := Y_1 Y_2^{-1}$, we have $Y \in \mathbf{RL}_\infty$ and

$$\begin{bmatrix} Y_1 \\ Y_2 \end{bmatrix} \mathbf{H}_2 = \mathbf{G}_Y \ . \tag{12}$$

From the first claim $\mathbf{G}_Y$ is negative, so from Lemma 2 $\|Y\|_\infty \leq 1$. Finally, (8) and (12) imply that $\mathbf{G}_Y = M \mathbf{G}_X$. □

Now we have the solution to the Nehari problem as posed in terms of S.

Theorem 1. The set of all matrices S in $\mathbf{RL}_\infty$ such that $\|S\|_\infty \leq 1$ and $R - S \in \mathbf{RH}_\infty$ is given by the formulas

$$S = X_1 X_2^{-1}$$

$$\begin{bmatrix} X_1 \\ X_2 \end{bmatrix} = L \begin{bmatrix} Y \\ I \end{bmatrix}$$

$$Y \in \mathbf{RH}_\infty , \quad \|Y\|_\infty \leq 1 .$$

Proof. First suppose

$$S \in \mathbf{RL}_\infty , \quad \|S\|_\infty \leq 1 , \quad R - S \in \mathbf{RH}_\infty .$$

By Lemma 2 and (5), $\mathbf{G}_S$ is negative and

$$\mathbf{G}_S \subset L (\mathbf{H}_2 \oplus \mathbf{H}_2) . \tag{13}$$

Define $M := L^{-1}$. From (13)

$$M \mathbf{G}_S \subset \mathbf{H}_2 \oplus \mathbf{H}_2 , \tag{14}$$

so (6) holds with S substituted for X. Also

$$\begin{aligned} L(\{0\} \oplus \mathbf{H}_2) &\subset L(\mathbf{H}_2 \oplus \mathbf{H}_2) \\ &= G(\mathbf{H}_2 \oplus \mathbf{H}_2) \text{ from (5)} \\ &= \begin{bmatrix} I & R \\ 0 & I \end{bmatrix} (\mathbf{H}_2 \oplus \mathbf{H}_2) \end{aligned}$$

$$\subset \mathbf{L}_2 \oplus \mathbf{H}_2 \,,$$

so (7) holds. Noting that M is J-unitary, invoke Lemma 3 to get the existence of Y in $\mathbf{RL}_\infty$ such that

$$\|Y\|_\infty \leq 1 \,, \quad \mathbf{G}_Y = M\,\mathbf{G}_S \,.$$

Since $\mathbf{G}_Y \subset \mathbf{H}_2 \oplus \mathbf{H}_2$ from (14), we have by Lemma 1 that actually $Y \in \mathbf{RH}_\infty$. Define

$$\begin{bmatrix} X_1 \\ X_2 \end{bmatrix} := L \begin{bmatrix} Y \\ I \end{bmatrix} ,$$

so that

$$\begin{bmatrix} X_1 \\ X_2 \end{bmatrix} \mathbf{H}_2 = L\,\mathbf{G}_Y$$

$$= \mathbf{G}_S$$

$$= \begin{bmatrix} S \\ I \end{bmatrix} \mathbf{H}_2 \,. \tag{15}$$

Pre-multiply (15) by $[0 \quad I]$ to get $X_2\mathbf{H}_2 = \mathbf{H}_2$. Thus $X_2^{-1} \in \mathbf{RH}_\infty$ by Lemma 1. Now pre-multiply (15) by $[I \quad -S]$ to get $S = X_1 X_2^{-1}$.

Conversely, suppose $Y \in \mathbf{RH}_\infty$ and $\|Y\|_\infty \leq 1$. Then

$$L\,\mathbf{G}_Y \subset L\,(\mathbf{H}_2 \oplus \mathbf{H}_2)$$

$$= G\,(\mathbf{H}_2 \oplus \mathbf{H}_2)$$

$$\subset \mathbf{L}_2 \oplus \mathbf{H}_2 \,,$$

and

$$\{0\} \oplus \mathbf{H}_2 = G\,(\{0\} \oplus \mathbf{H}_2)$$

$$\subset G\,(\mathbf{H}_2 \oplus \mathbf{H}_2)$$

$$= L\,(\mathbf{H}_2 \oplus \mathbf{H}_2)$$

$$\subset L\,(\mathbf{L}_2 \oplus \mathbf{H}_2) \,.$$

Invoke Lemma 3 again: there exists S in $\mathbf{RL}_\infty$ such that $\|S\|_\infty \leq 1$ and

$\mathbf{G}_S = L\,\mathbf{G}_Y$. Define

$$\begin{bmatrix} X_1 \\ X_2 \end{bmatrix} := L \begin{bmatrix} Y \\ I \end{bmatrix}.$$

Then

$$\begin{bmatrix} X_1 \\ X_2 \end{bmatrix} \mathbf{H}_2 = \mathbf{G}_S ,$$

so $S = X_1 X_2^{-1}$ as before. □

The formulas in Theorem 1 yield S as a linear fractional transformation of Y. To see this, partition L as

$$L = \begin{bmatrix} L_1 & L_2 \\ L_3 & L_4 \end{bmatrix}.$$

Then

$$S = (L_1 Y + L_2)(L_3 Y + L_4)^{-1} .$$

One possible candidate for Y is $Y = 0$, in which case S is simply $L_2 L_4^{-1}$.

Let's summarize the results of this section in the form of an algorithm. The input is a matrix R in $\mathbf{RL}_\infty$ having the properties

R is strictly proper

$R^\sim \in \mathbf{RH}_\infty$

$\|\Gamma_R\| < 1$

and the output is a matrix X in $\mathbf{RH}_\infty$ such that $\|R - X\|_\infty \leq 1$.

Step 1. Find a minimal realization of R:

$$R(s) = [A, B, C, 0] .$$

Step 2. Solve the Lyapunov equations

$$AL_c + L_c A^T = BB^T$$

$$A^T L_o + L_o A = C^T C$$

and set $N := (I - L_o L_c)^{-1}$.

Step 3. Set

$$L_1(s) = [A, -L_c NC^T, C, I]$$

$$L_2(s) = [A, N^T B, C, 0]$$

$$L_3(s) = [-A^T, NC^T, -B^T, 0]$$

$$L_4(s) = [-A^T, NL_o B, B^T, I].$$

Step 4. Select Y in $\mathbf{RH}_\infty$ with $\|Y\|_\infty \le 1$ (for example $Y=0$) and set

$$X = R - (L_1 Y + L_2)(L_3 Y + L_4)^{-1}.$$

Exercise 1. Derive the formulas for L_i in Step 3 using the equations

$$L = GG_-^{-1},$$

$$G(s) = \left[A, [0 \ \ B], \begin{bmatrix} C \\ 0 \end{bmatrix}, I\right],$$

and (7.5.6).

Example 1.

Let

$$R(s) = \begin{bmatrix} \frac{.5}{s-1} & 0 \\ \frac{1}{s^2-s+1} & \frac{2}{s-1} \end{bmatrix}.$$

We computed in Section 5.2 that dist$(R, \mathbf{RH}_\infty) = 1.2695$. Scale R by, say, $1/1.28$, i.e. redefine R to be

$$R(s) = \frac{1}{1.28} \begin{bmatrix} \frac{.5}{s-1} & 0 \\ \frac{1}{s^2-s+1} & \frac{2}{s-1} \end{bmatrix}.$$

The algorithm applied to this matrix has the following results:

Step 1.

$$R(s) = [A, B, C, D]$$

$$A = \begin{bmatrix} 2 & -2 & 1 & 0 \\ 1 & 0 & 0 & 0 \\ 0 & 1 & 0 & 0 \\ 0 & 0 & 0 & 1 \end{bmatrix}, \quad B = \begin{bmatrix} .7813 & 0 \\ 0 & 0 \\ 0 & 0 \\ 0 & 1.563 \end{bmatrix}$$

$$C = \begin{bmatrix} .5 & -.5 & .5 & 0 \\ 0 & 1 & -1 & 1 \end{bmatrix}, \quad D = 0$$

Step 2.

$$L_c = \begin{bmatrix} .2035 & 0 & -.1017 & 0 \\ 0 & .1017 & 0 & 0 \\ -.1017 & 0 & .2035 & 0 \\ 0 & 0 & 0 & 1.221 \end{bmatrix}$$

$$L_o = \begin{bmatrix} .6250 & -1.125 & .6250 & -.3333 \\ -1.125 & 2.625 & -1.625 & 1 \\ .6250 & -1.625 & 1.125 & -.6667 \\ -.3333 & 1 & -.6667 & .5 \end{bmatrix}$$

$$N = \begin{bmatrix} 1.507 & -5.447 & 5.093 & -29.27 \\ -1.253 & 15.69 & -13.83 & 79.37 \\ .7918 & -9.640 & 10.11 & -52.14 \\ -.5339 & 6.614 & -6.251 & 36.90 \end{bmatrix}$$

Step 4. Take $Y = 0$. Then

$$X(s) = R(s) - L_2(s) L_4(s)^{-1}$$

$$= [A, B, C, 0] - [A, N^T B, C, 0] \times [-A^T, NL_o B, B^T, I]^{-1}$$

$$= \left[\begin{bmatrix} A & 0 & 0 \\ 0 & A & -N^T BB^T \\ 0 & 0 & -A^T - NL_o BB^T \end{bmatrix}, \begin{bmatrix} B \\ N^T B \\ NL_o B \end{bmatrix}, [C \;\; -C \;\; 0], 0 \right].$$

The resulting $X(s)$ is

$$\begin{bmatrix} \dfrac{-4.315(s+2.473)(s^2+1.206s+.3951)}{(s+84.53)(s+1.101)(s^2+1.381s+1.110)} & \dfrac{10.47(s+1.054)(s+1)^2}{(s+84.53)(s+1.101)(s^2+1.381s+1.110)} \\ \dfrac{31.10(s+1.086)(s^2+1.118s+.7120)}{(s+84.53)(s+1.101)(s^2+1.381s+1.110)} & \dfrac{-76.19(s^2+1.279s+.9036)}{(s+84.53)(s^2+1.381s+1.110)} \end{bmatrix}.$$

Exercise 2. Take

$$T_1(s) = \frac{1}{s+4}, \quad T_2(s) = \frac{s-1}{s+1}, \quad T_3(s) = 1 .$$

First, compute α using the method of Section 6.2. Second, choose $\gamma > \alpha$ (keep γ variable) and define R as in Theorem 8.1.1. Third, using the previous algorithm find an X in $\mathbf{RH}_\infty$ such that $\|R - X\|_\infty \leq 1$. Fourth, get Q from Theorem 8.1.1. This Q depends on γ. Finally, in the coefficients of Q let γ tend to α. You should get the optimal Q for the original model-matching problem.

8.4 Summary: Solution of the Standard Problem

We have completed our solution of the standard problem posed in Chapter 3. The first step is to reduce the standard problem to a model-matching problem; state-space tools for this reduction are given in Section 4.5. In the single-input, single-output case the model-matching problem is relatively easy to solve (Section 6.2). In the multi-input, multi-output case the model-matching problem is reduced to the Nehari problem (Section 8.1). Finally, the Nehari problem is solved in Section 8.3.

Notes and References

The first multivariable $\mathbf{H}_\infty$ problem to be solved was the disturbance attenuation problem (equivalently, the weighted sensitivity problem): Chang and Pearson (1984) used matrix interpolation theory, Francis, Helton, and Zames (1984) used the geometric Ball-Helton theory, and Safonov and Verma (1985) used operator theory.

The approach of Section 1, called γ-iteration, is due to Doyle (1984); Francis (1983) independently developed a similar approach for the case $T_3 = I$. A detailed treatment of γ-iteration is contained in Chu, Doyle, and Lee (1986).

For a comprehensive treatment of Krein spaces see Bognar (1974). The approach in Section 3 to the Nehari problem is due to Ball and Helton (1983) and Ball and Ran (1986); the proofs are modifications of those in Francis *et al.* (1984). An alternative state-space approach to the Nehari problem (which inspired the work of Ball and Ran) is that of Glover (1984).

An alternative approach to the model-matching problem is that of Kwakernaak (1985, 1986).

In the scalar-valued model-matching problem the value of α can be computed directly: it's the norm of a certain Hankel operator. The method presented in this chapter for computing α in the matrix-valued case is iterative: α isn't equal to the norm of a Hankel operator except in the very special case where T_2 and T_3 are both square and nonsingular. Feintuch and Francis (1986) showed that α equals the norm of a certain (non-Hankel) operator. An alternative formula has been derived by Young (1986b). This latter formula is simple enough to state here. Define two subspaces $\mathbf{X}$ and $\mathbf{Y}$ of $\mathbf{L}_2$,

$$\mathbf{X} := T_3^{-1}\mathbf{H}_2$$

$$:= \{f \in \mathbf{L}_2 : T_3 f \in \mathbf{H}_2\}$$

$$\mathbf{Y} := \text{orthogonal complement of } T_2\mathbf{H}_2 \text{ in } \mathbf{L}_2 .$$

Now define the operator Ξ from $\mathbf{X}$ to $\mathbf{Y}$ as follows:

$$\Xi f := \text{orthogonal projection of } T_1 f \text{ onto } \mathbf{Y}, \quad f \in \mathbf{X} .$$

Then Young's formula is $\alpha = ||\Xi||$. Along these lines, an alternative approach to computing α is that of Jonckheere and Juang (1986).

CHAPTER 9

PERFORMANCE BOUNDS

For some simple examples of the standard problem it's possible to obtain useful bounds on achievable performance, sometimes even to characterize achievable performance exactly. This brief chapter presents three illustrative examples.

Figure 1 shows a feedback system with a disturbance signal w referred to the output of the plant P. As usual, P is strictly proper and K is proper. The transfer matrix from w to y is the sensitivity matrix $S := (I - PK)^{-1}$.

Suppose first that the spectrum of w is confined to a pre-specified interval of frequencies $[0,\omega_1]$, $\omega_1 > 0$. Then the problem of attenuating the effect of w on the output y of the plant is equivalent to that of making $||S(j\omega)||$ uniformly small on the interval $[0,\omega_1]$. As in Section 6.3 introduce the characteristic function

$$\chi(j\omega) := 1 \text{ if } |\omega| \le \omega_1$$

$$:= 0 \text{ if } |\omega| > \omega_1 .$$

Then the maximum value of $||S(j\omega)||$ over the interval $[0,\omega_1]$ equals the $\mathbf{L}_\infty$-norm of χS. It may happen that as we try to make $||\chi S||_\infty$ smaller and smaller, the global bound $||S||_\infty$ becomes larger and larger. This is unpleasant because a large value of $||S||_\infty$ means the system has poor stability margin. Think of the scalar-valued case: if $||S||_\infty$ is large, then $|1-(PK)(j\omega)|$ is small at some frequency, i.e. the Nyquist plot of PK passes near the critical point $s=1$.

The first result says that if P is minimum phase (in a certain sense), then $||\chi S||_\infty$ can be made as small as desired while $||S||_\infty$ is simultaneously maintained less than any bound δ. Of course, δ must be greater than 1 since $||S||_\infty \ge 1$ for every stabilizing K.

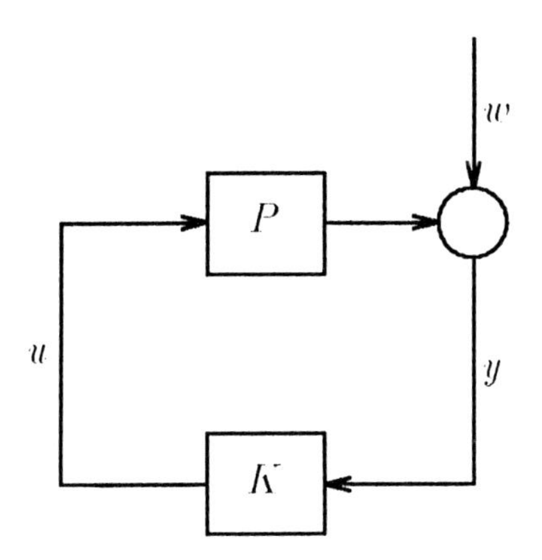

Figure 9.1. Disturbance rejection

Theorem 1. If P has a right-inverse which is analytic in Re $s \geq 0$, then for every $\epsilon > 0$ and $\delta > 1$ there exists a stabilizing K such that

$$\|\chi S\|_\infty < \epsilon, \quad \|S\|_\infty < \delta .$$

Proof. The idea is to approximately invert P over the frequency range $[0, \omega_1]$ while rolling off fast enough at higher frequencies. The first step is to parametrize all stabilizing K's as in Section 4.4. Bring in a doubly-coprime factorization of P:

$$P = NM^{-1} = \tilde{M}^{-1}\tilde{N} \tag{1}$$

$$\begin{bmatrix} \tilde{X} & -\tilde{Y} \\ -\tilde{N} & \tilde{M} \end{bmatrix} \begin{bmatrix} M & Y \\ N & X \end{bmatrix} = I . \tag{2}$$

Then the formula for K is (Theorem 4.4.1)

$$\begin{aligned} K &= (Y - MQ)(X - NQ)^{-1} \\ &= (\tilde{X} - Q\tilde{N})^{-1}(\tilde{Y} - Q\tilde{M}) \end{aligned} \tag{3}$$

$$Q \in \mathbf{RH}_\infty .$$

With these two representations of P and K and using (2) we get

$$S = (X - NQ)\tilde{M} . \tag{4}$$

Now fix $\epsilon > 0$ and $\delta > 1$. Choose $c > 0$ so small that

$$c\,\|X\tilde{M}\|_\infty < \min(\epsilon, \delta) \tag{5}$$

$$(1+c)^2 < \delta . \tag{6}$$

It follows from (2) that

$$X\tilde{M} - N\tilde{Y} = I .$$

Since P is strictly proper, so is N. Hence

$$X(\infty)\tilde{M}(\infty) = I ,$$

so that

$$\|X(\infty)\tilde{M}(\infty)\| = 1 .$$

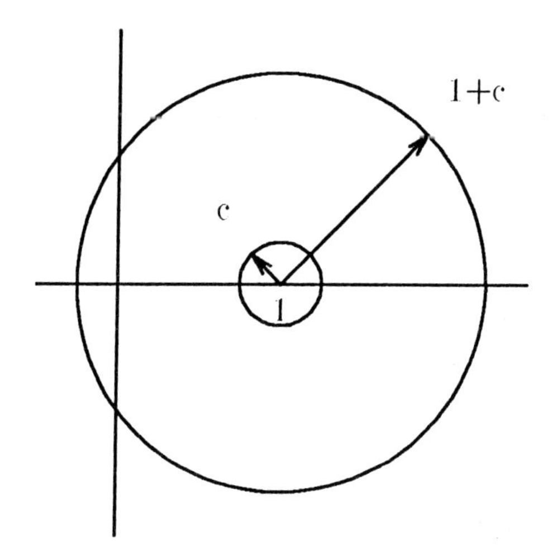

Figure 9.2. For proof of Theorem 1

Since $||X(j\omega)\tilde{M}(j\omega)||$ is a continuous function of ω, it's possible to choose $\omega_2 \geq \omega_1$ such that

$$||X(j\omega)\tilde{M}(j\omega)|| \leq 1 + c \ , \ \ \omega \geq \omega_2 \ . \tag{7}$$

Bring in another characteristic function,

$$\begin{aligned} \chi_2(j\omega) &:= 1 \ , \quad |\omega| \leq \omega_2 \\ &:= 0 \ , \quad |\omega| > \omega_2 \ , \end{aligned}$$

so that (7) is equivalent to

$$||(1-\chi_2)X\tilde{M}||_\infty \leq 1 + c \ . \tag{8}$$

The assumption on P implies that N has a right-inverse, say N_{ri}, which is stable (but not proper). Choose a scalar-valued function V in $\mathbf{RH}_\infty$ with the following three properties:

VN_{ri} is proper

$$||\chi_2(1-V)||_\infty \leq c \tag{9}$$

$$||1-V||_\infty \leq 1 + c \ . \tag{10}$$

The idea behind the choice of V can be explained by the picture of the complex plane in Figure 2. The Nyquist plot of V should lie in the smaller disk up to frequency ω_2 (inequality (9)), and in the larger disk thereafter (inequality (10)). In addition, V should roll off fast enough so that VN_{ri} is proper. For example, V could take the form

$$V(s) = \frac{1}{(\tau s + 1)^k}$$

for large enough τ and k.

Finally, take Q to be

$$Q := VN_{ri}X \ .$$

Substitution into (4) gives

$$\begin{aligned} S &= (X - VNN_{ri}X)\tilde{M} \\ &= (1-V)X\tilde{M} \ . \end{aligned}$$

Thus

$$\begin{aligned}\|\chi S\|_\infty &\leq \|\chi_2 S\|_\infty \text{ since } \omega_2 \geq \omega_1 \\ &= \|\chi_2(1-V)X\tilde{M}\|_\infty \\ &\leq \|\chi_2(1-V)\|_\infty \|X\tilde{M}\|_\infty \\ &\leq c\,\|X\tilde{M}\|_\infty \text{ from (9)} \\ &< \min(\epsilon, \delta) \text{ from (5)}\end{aligned}$$

and

$$\begin{aligned}\|(1-\chi_2)S\|_\infty &= \|(1-\chi_2)(1-V)X\tilde{M}\|_\infty \\ &\leq \|1-V\|_\infty \|(1-\chi_2)X\tilde{M}\|_\infty \\ &\leq (1+c)^2 \text{ from (8) and (10)} \\ &< \delta \text{ from (6)}\,.\end{aligned}$$

The two inequalities

$$\|\chi_2 S\|_\infty < \delta$$

$$\|(1-\chi_2)S\|_\infty < \delta$$

imply $\|S\|_\infty < \delta$. □

On the other hand, if P has a zero in the right half-plane, then $\|S\|_\infty$ must necessarily increase without limit if $\|\chi S\|_\infty$ tends to zero. This might be described as the "waterbed effect".

Theorem 2. Assume there is a point s_0 in Re $s > 0$ such that the rank of $P(s_0)$ is less than the number of its rows. Then there exists a positive real number a such that for every stabilizing K

$$\|\chi S\|_\infty \|S\|_\infty^a \geq 1\,.$$

Proof. We continue with the notation of the previous proof. Fix some stabilizing K. Then S must have the form

$$S = (X - NQ)\tilde{M} \ . \tag{4bis}$$

By assumption, there is a point s_0 in Re $s > 0$ and a nonzero complex vector x such that

$$x^* P(s_0) = 0 \ . \tag{11}$$

Scale x so that $x^* x = 1$.

It is claimed that

$$x^* S(s_0) = x^* \ . \tag{12}$$

To see this, first get from (2) that

$$\tilde{M}X - \tilde{N}Y = I \ .$$

Pre-multiply by $\tilde{M}^{-1}$, post-multiply by $\tilde{M}$, and use (1) to get

$$X\tilde{M} - PY\tilde{M} = I \ .$$

Therefore from (11)

$$x^* X(s_0)\tilde{M}(s_0) = x^* \ . \tag{13}$$

Similarly, (11) also implies that

$$x^* N(s_0) = 0 \ . \tag{14}$$

Then (13) and (14) imply (12).

Map the right half-plane onto the unit disk via the mapping

$$s \rightarrow z = \frac{s_0 - s}{\bar{s}_0 + s}$$

$$z \rightarrow s = \frac{s_0 - \bar{s}_0 z}{1+z} \ .$$

The point $s = s_0$ is mapped to the origin, $z = 0$, and the interval $[0, j\omega_1]$ is mapped onto an arc

$$\{e^{j\theta} : \theta_1 \leq \theta \leq \theta_2\} \ . \tag{15}$$

Let ϕ be the angle subtended by this arc, i.e. $\phi = |\theta_2 - \theta_1|$, and define

$$R(z) := S[(s_0 - \bar{s}_0 z)/(1+z)] \ . \tag{16}$$

Then R is rational and analytic in the closed unit disk. Moreover

$$\max\{||R(e^{j\theta})||:\theta_1\leq\theta\leq\theta_2\} = ||\chi S||_\infty \tag{17}$$

and

$$\max\{||R(e^{j\theta})||:0\leq\theta\leq 2\pi\} = ||S||_\infty . \tag{18}$$

Now let n be any integer greater than $2\pi/\phi$ and define

$$T(z) := R(z)R(ze^{j2\pi/n})...R(ze^{j2\pi(n-1)/n}) . \tag{19}$$

Then T is rational and analytic in the closed unit disk. Since the angle $2\pi/n$ is less than ϕ, at least one of the n points

$$ze^{j2\pi k/n}, \quad k=0,...,n-1$$

lies in the arc (15) for each z on the unit circle. Thus from (19)

$$\begin{aligned} c &:= \max\{||T(e^{j\theta})||:0\leq\theta\leq 2\pi\} \\ &\leq [\max\{||R(e^{j\theta})||:0\leq\theta\leq 2\pi\}]^{n-1} \\ &\times [\max\{||R(e^{j\theta})||:\theta_1\leq\theta\leq\theta_2\}] , \end{aligned}$$

or from (17) and (18)

$$c \leq ||S||_\infty^a ||\chi S||_\infty , \tag{20}$$

where $a:=n-1$. Notice that n depends only on ϕ, which in turn depends only on s_0 and ω_1. Thus a is independent of K. It remains to show that $c\geq 1$.

Now (12) and (16) imply

$$x^*R(0) = x^* .$$

This and (19) yield

$$x^*T(0) = x^* .$$

Thus

$$x^*T(0)x = x^*x = 1 . \tag{21}$$

But $x^*T(z)x$ is analytic in the closed unit disk. So (21) and the maximum modulus theorem imply that

$$|x^*T(e^{j\theta_0})x| \geq 1 \tag{22}$$

for some θ_0. Now

$$c^2 = \max_{\theta} \max_{y^* y = 1} y^* T(e^{j\theta}) T(e^{j\theta})^* y$$

$$\geq x^* T(e^{j\theta_0}) T(e^{j\theta_0})^* x \ . \tag{23}$$

Define the vector

$$x_0 := T(e^{j\theta_0})^* x \ .$$

Then

$$1 \leq | x_0^* x | \text{ from (22)}$$

$$\leq (x_0^* x_0)^{1/2} (x^* x)^{1/2}$$

$$= (x_0^* x_0)^{1/2}$$

$$\leq c \text{ from (23)} . \ \Box$$

For the third result consider, again with respect to Figure 1, the problem of attenuating the effect of w (no longer restricted to be bandlimited) on the control signal u; that is, the problem is to achieve feedback stability by a controller which limits as much as possible the control effort. The transfer matrix from w to u equals KS, so the objective is to minimize $||KS||_\infty$. The case where P is stable is trivial: an optimal K is $K = \mathbf{0}$. So we suppose K is not stable. For technical reasons it is assumed that P has no poles on the imaginary axis; thus P belongs to $\mathbf{RL}_\infty$ but not $\mathbf{RH}_\infty$.

Bring in Γ_P, the Hankel operator with symbol P. Let $\sigma_{\min}(\Gamma_P)$ denote its smallest singular value, i.e. the square root of the smallest nonzero eigenvalue of $\Gamma_P^* \Gamma_P$. This number can be easily computed via Theorem 5.1.3.

Theorem 3. If P belongs to $\mathbf{RL}_\infty$ but not $\mathbf{RH}_\infty$, then the minimum value of $||KS||_\infty$ over all stabilizing K's equals the reciprocal of $\sigma_{\min}(\Gamma_P)$.

Proof. The proof is an interesting application of operator theory; some of the details are left as exercises.

Again, we continue with the notation introduced in the proof of Theorem 1. From (1) and (2) we have

$$P = NM^{-1}$$

$$\tilde{X}M - \tilde{Y}N = I ,$$

so that

$$\tilde{X} - \tilde{Y}P = M^{-1} .$$

Thus $M^{-1} \in \mathbf{RL}_\infty$. It follows from Theorem 7.4.1 that M has an inner-outer factorization:

$$M = M_i M_o ,$$

$$M_i \text{ inner} ,$$

$$M_o , M_o^{-1} \in \mathbf{RH}_\infty .$$

Thus $P = (NM_o^{-1})M_i^{-1}$. Hence we may as well assume from the start that M is inner; similarly that $\tilde{M}$ is co-inner.

From (3) and (4) we have

$$KS = (Y - MQ)\tilde{M} .$$

Hence the minimum value of $\|KS\|_\infty$ over all stabilizing K's equals

$$\min\{\|(Y - MQ)\tilde{M}\|_\infty : Q \in \mathbf{RH}_\infty\}$$

$$= \min\{\|Y - MQ\|_\infty : Q \in \mathbf{RH}_\infty\}$$

$$= \text{dist}(M^{-1}Y, \mathbf{RH}_\infty)$$

$$= \text{dist}(M^{-1}Y, \mathbf{H}_\infty) \text{ by Lemma 8.1.3}$$

$$= \|\Gamma_R\| \text{ by Nehari's theorem,}$$

where

$$R := M^{-1}Y . \tag{24}$$

So proving the theorem is equivalent to showing that

$$\|\Gamma_R\| = [\sigma_{\min}(\Gamma_P)]^{-1} . \tag{25}$$

It would simplify matters if the two Hankel operators in (25) were surjective, but they're not: they're finite rank (Theorem 5.1.1). To make them surjective, we'll restrict their co-domains.

The image of Γ_P is $\Pi_1 P\,\mathbf{H}_2$. This subspace of $\mathbf{H}_2^{\perp}$ is finite-dimensional, hence closed. Let's redefine Γ_P to be

$$\Gamma_P : \mathbf{H}_2 \to \Pi_1 P\,\mathbf{H}_2$$

$$\Gamma_P f \ := \Pi_1 P f \ .$$

Now $P = NM^{-1}$ and M^{-1} is the symbol of the Hankel operator

$$\Gamma_{M^{-1}} : \mathbf{H}_2 \to \Pi_1 M^{-1}\mathbf{H}_2 \ .$$

Introduce a new operator,

$$\Omega_N : \Pi_1 M^{-1}\mathbf{H}_2 \to \Pi_1 P\,\mathbf{H}_2$$

$$\Omega_N f \ := \Pi_1 N f \ .$$

Exercise 1. Verify that

$$\Gamma_P = \Omega_N \Gamma_{M^{-1}} \ . \tag{26}$$

It follows from (2) that $M\tilde{Y} = Y\tilde{M}$. Thus $R = \tilde{Y}\tilde{M}^{-1}$.

Exercise 2. Use the fact that M is inner to prove

$$\Gamma_{M^{-1}} \Gamma^*_{M^{-1}} = I \ . \tag{27}$$

Similarly

$$\Gamma_{\tilde{M}^{-1}} \Gamma^*_{\tilde{M}^{-1}} = I \ . \tag{28}$$

It follows from (2) that

$$X\tilde{M} - N\tilde{Y} = I$$

$$\tilde{X}M - \tilde{Y}N = I \ .$$

Exercise 3. Use these two equations to prove

$$\Omega_N \Omega_{\tilde{Y}} = -I \ , \quad \Omega_{\tilde{Y}} \Omega_N = -I \ .$$

These two equations imply that

$$\Omega_{\tilde{Y}} \Omega_{\tilde{Y}}^{*} = (\Omega_N^{*} \Omega_N) . \tag{29}$$

From (26) and (27) we have

$$\Gamma_P \Gamma_P^{*} = \Omega_N \Omega_N^{*} . \tag{30}$$

Similarly

$$\Gamma_R \Gamma_R^{*} = \Omega_{\tilde{Y}} \Omega_{\tilde{Y}}^{*} . \tag{31}$$

Thus

$$\begin{aligned} \|\Gamma_R \Gamma_R^{*}\| &= \|\Omega_{\tilde{Y}} \Omega_{\tilde{Y}}^{*}\| \text{ from (31)} \\ &= \|(\Omega_N^{*} \Omega_N)^{-1}\| \text{ from (29)} \\ &= \|(\Omega_N \Omega_N^{*})^{-1}\| \\ &= \|(\Gamma_P \Gamma_P^{*})^{-1}\| \text{ from (30).} \end{aligned}$$

This proves (25) since the largest eigenvalue of $(\Gamma_P \Gamma_P^{*})^{-1}$ equals the reciprocal of the smallest eigenvalue of $\Gamma_P \Gamma_P^{*}$. □

Example 1.

The simplest possible example to illustrate Theorem 3 is

$$P(s) = \frac{1}{s-1} .$$

Then

$$\sigma_{\min}(\Gamma_P) = \|\Gamma_P\| = 1/2 .$$

Let's stabilize with a constant controller, $K(s) = -k$. Then

$$(KS)(s) = -k \; \frac{s-1}{s+k-1} .$$

Clearly K stabilizes P iff $k > 1$. We compute that

$$\begin{aligned} \|KS\|_\infty &= k \text{ if } k \geq 2 \\ &= \frac{k}{k-1} \text{ if } 1 < k < 2 . \end{aligned}$$

Thus $||KS||_\infty$ is minimized by the gain $k=2$.

Notes and References

The idea of Theorem 1 is due to Bensoussan and Zames: Bensoussan (1984) proved the theorem in the scalar-valued case with P stable, and Zames and Bensoussan (1983) proved a result like Theorem 1 but where P was assumed to be diagonally dominant at high frequency and K was required to be diagonal. The proof given here is from Francis (1983). Theorem 2 is the multivariable generalization of a result of Francis and Zames (1984). Theorem 3 is due independently to Glover (1986), whose proof uses state-space methods, in contrast to the operator-theoretic one here, and to Verma (1985). Actually, their result pertains to the mathematically equivalent problem of robust stabilization: the largest radius of plant uncertainty for robust stabilizability equals the reciprocal of the smallest Hankel singular value of the nominal plant.

For other results on achievable performance see, for example, Boyd and Desoer (1984), Freudenberg and Looze (1985), and O'Young and Francis (1985, 1986).

BIBLIOGRAPHY

Adamjan, V.M., D.Z. Arov, and M.G. Krein (1971). "Analytic properties of Schmidt pairs for a Hankel operator and the generalized Schur-Takagi problem," *Math. USSR Sbornik*, vol. 15, pp. 31-73.

Adamjan, V.M., D.Z. Arov, and M.G. Krein (1978). "Infinite block Hankel matrices and related extension problems," *AMS Transl.*, vol. 111, pp. 133-156.

Allison, A.C. and N.J. Young (1983). "Numerical algorithms for the Nevanlinna-Pick problem," *Numerische Mathematik*, vol. 42, pp. 125-145.

Ball, J.A. and J.W. Helton (1983). "A Beurling-Lax theorem for the Lie group U(m,n) which contains most classical interpolation theory," *J. Op. Theory*, vol. 9, pp. 107-142.

Ball, J.A. and A.C.M. Ran (1986). "Optimal Hankel norm model reductions and Wiener-Hopf factorizations I: the canonical case," *SIAM J. Control and Opt.* To appear.

Bart, H., I. Gohberg, and M.A. Kaashoek (1979). *Minimal Factorization of Matrix and Operator Functions,* Birkhauser, Basel.

Bensoussan, D. (1984). "Sensitivity reduction in single-input single-output systems," *Int. J. Control*, vol. 39, pp. 321-335.

Bognar, J. (1974). *Indefinite Inner Product Spaces,* Springer-Verlag, New York.

Boyd, S. and C.A. Desoer (1984). "Subharmonic functions and performance bounds on linear time-invariant feedback systems," Memo. No. UCB/ERL M84/51, Elect. Res. Lab., Univ. of Calif. , Berkeley, CA.

Chang, B.C. and J.B. Pearson (1984). "Optimal disturbance reduction in linear multivariable systems," *IEEE Trans. Auto. Cont.*, vol. AC-29, pp. 880-887.

Chen, B.S. (1984). "Controller synthesis of optimal sensitivity: multivariable case," *Proc. IEE*, vol. 131.

Chen, M.J. and C.A. Desoer (1982). "Necessary and sufficient condition for robust stability of linear distributed feedback systems," *Int. J. Cont.*, vol. 35, pp. 255-267.

Cheng, L. and J.B. Pearson (1978). "Frequency domain synthesis of multivariable linear regulators," *IEEE Trans. Auto. Cont.*, vol. AC-23, pp. 3-15.

Chu, C.C. (1985). "H_∞ Optimization and robust multivariable control," Ph. D. Thesis, Dept. Elect. Eng., Univ. Minn., Minneapolis, MN.

Chu, C.C., J.C. Doyle, and E.B. Lee (1986). "The general distance problem in H_∞ optimal control theory," *Int. J. Control*, vol. 44, pp. 565-596.

Conway, J.B. (1985). *A Course in Functional Analysis,* Springer-Verlag, New York.

Curtain, R.F. and K. Glover (1984). "Robust stabilization of infinite-dimensional systems by finite-dimensional controllers: derivations and examples," *Proc. Sym. MTNS*, Stockholm.

Curtain, R.F. and K. Glover (1986). "Robust stabilization of infinite-dimensional systems by finite-dimensional controllers," *Systems and Control Letters*, vol. 7, pp. 41-48.

Delsarte, P., Y. Genin, and Y. Kamp (1979). "The Nevanlinna-Pick problem for matrix-valued functions," *SIAM J. Appl. Math.*, vol. 36, pp. 47-61.

Desoer, C.A. and M. Vidyasagar (1975). *Feedback Systems: Input-Output Properties,* Academic Press, New York.

Desoer, C.A., R.W. Liu, J. Murray, and R. Saeks (1980). "Feedback system design: the fractional representation approach," *IEEE Trans. Auto. Cont.*, vol. AC-25, pp. 399-412.

Doyle, J.C. and G. Stein (1981). "Multivariable feedback design: concepts for a classical modern synthesis," *IEEE Trans. Auto. Cont.*, vol. AC-26, pp. 4-16.

Doyle, J.C. (1983). "Synthesis of robust controllers and filters," *Proc. CDC.*

Doyle, J.C. (1984). "Lecture Notes in Advances in Multivariable Control," *ONR/Honeywell Workshop*, Minneapolis, MN.

Doyle, J.C. and C.C. Chu (1985). "Matrix interpolation and H_∞ performance bounds," *Proc. ACC.*

Doyle, J.C. (1985a). "Robust stability with structured perturbations," *Proc. Sym. on MTNS*, Stockholm.

Doyle, J.C. (1985b). "Structured uncertainty in control systems," *IFAC Workshop on Model Error Concepts and Compensation*, Boston, MA.

Duren, P.L. (1970). *Theory of H_p Spaces,* Academic Press, New York.

Dym, H. and H.P. McKean (1972). *Fourier Series and Integrals,* Academic Press, New York.

Feintuch, A. and B.A. Francis (1984). "Uniformly optimal control of linear time-varying systems," *Systems and Control Letters*, vol. 5, pp. 67-71.

Feintuch, A., P. Khargonekar, and A. Tannenbaum (1986). "On the sensitivity minimization problem for linear time-varying systems," *SIAM J. Control and Opt.* To appear.

Feintuch, A. and A. Tannenbaum (1986). "Gain optimization for distributed systems," *Systems and Control Letters*, vol. 6, pp. 295-302.

Feintuch, A. and B.A. Francis (1986). "Uniformly optimal control of linear systems," *Automatica*, vol. 21, pp. 563-574.

Flamm, D.S. and S.K. Mitter (1985). "Progress on H_∞ optimal sensitivity for delay systems," Tech. Rept. LIDS-P-1513, MIT, Cambridge, MA.

Flamm, D.S. (1986). "Control of delay systems for minimax sensitivity," Tech. Rept. LIDS-TH-1560, MIT, Cambridge, MA.

Foias, C. and A. Tannenbaum (1986). "On the Nehari problem for a certain class of L_∞ functions appearing in control theory," *J. Funct. Analysis.* To appear.

Foias, C., A. Tannenbaum, and G. Zames (1986). "Weighted sensitivity minimization for delay systems," *SIAM J. Control and Opt.* To appear.

Foo, Y.K. and I. Postlethwaite (1984). "An H_∞-minimax approach to the design of robust control systems," *Systems and Control Letters*, vol. 5, pp. 81-88.

Francis, B.A. (1983). *Notes on H_∞-optimal linear feedback systems.* Lectures given at Linkoping Univ.

Francis, B.A. and G. Zames (1983). "Design of H_∞-optimal multivariable feedback systems," *Proc. CDC.*

Francis, B.A. and G. Zames (1984). "On H_∞-optimal sensitivity theory for siso feedback systems," *IEEE Trans. Auto. Cont.*, vol. AC-29, pp. 9-16.

Francis, B.A., J.W. Helton, and G. Zames (1984). "H_∞-optimal feedback controllers for linear multivariable systems," *IEEE Trans. Auto. Cont.*, vol. AC-29, pp. 888-900.

Francis, B.A. (1985). "Optimal disturbance attenuation with control weighting," in *Lecture Notes in Control and Inf. Sci., Vol. 66*, Springer-Verlag. Proc. 1984 Twente Workshop on Systems and Optimization.

Francis, B.A. and J.C. Doyle (1986). "Linear control theory with an H_∞ optimality criterion," *SIAM J. Control Opt.* To appear.

Freudenberg, J. (1985). "Issues in frequency domain feedback control," Ph. D. Thesis, Dept. Elect. Eng., Univ. Illinois, Urbana, Il.

Freudenberg, J. and D. Looze (1985). "Right half-plane poles and zeros and design trade-offs in feedback systems," *IEEE Trans. Auto. Cont.*, vol. AC-30, pp. 555-565.

Freudenberg, J. and D. P. Looze (1986). "An analysis of H_∞-optimization design methods," *IEEE Trans. Aut. Cont.*, vol. AC-31, pp. 194-200.

Ganesh, C. and J.B. Pearson (1985). "Design of optimal control systems with stable feedback," Tech. Rept. #8514, Dept. EE, Rice Univ., Houston.

Garnett, J.B. (1981). *Bounded Analytic Functions,* Academic Press, New York.

Georgiou, T. and P. Khargonekar (1986). "A constructive algorithm for sensitivity optimization of periodic systems," *SIAM J. Control and Opt.* To appear.

Glover, K. (1984). "All optimal Hankel norm approximations of linear multivariable systems and their L_∞-error bounds," *Int. J. Cont.*, vol. 39, pp. 1115-1193.

Glover, K. (1986). "Robust stabilization of linear multivariable systems: relations to approximation," *Int. J. Control*, vol. 43, pp. 741-766.

Gohberg, I. and S. Goldberg (1981). *Basic Operator Theory,* Birkhauser, Boston.

Golub, G.H. and C.F. Van Loan (1983). *Matrix Computations,* Johns Hopkins Univ. Press, Baltimore.

Grimble, M.J. (1984). "Optimal H_∞ robustness and the relationship to LQG design problems," Tech. Rept. ICU/55, ICU, Univ. Strathclyde, Glasgow.

Halmos, P.R. (1982). *A Hilbert Space Problem Book,* Springer-Verlag, New York.

Helton, J.W. (1976). "Operator theory and broadband matching," *Proc. Allerton Conf.*

Helton, J.W. and D.F. Schwartz (1985). "A primer on the H_∞ disk method in frequency-domain design: control," Tech. Report, Dept. Math., Univ. of Calif., San Diego.

Helton, J.W. (1985a). "Worst case analysis in the frequency-domain: an H_∞ approach to control," *IEEE Trans. Auto. Cont.*, vol. AC-30, pp. 1154-1170.

Helton, J.W. (1985b). *Lecture notes,* NSF-CBMS Conf. on Optimization in Operator Theory, Analytic Function Theory, and Electrical Engineering, Lincoln, NB.

Hoffman, K. (1962). *Banach Spaces of Analytic Functions,* Prentice-Hall, Englewood Cliffs, N.J.

Jonckheere, E. and M. Verma (1986). "A spectral characterization of H_∞-optimal feedback performance: the multivariable case," Tech. Rept., Dept. EE, USC, Los Angeles.

Jonckheere, E.A., M.G. Safonov, and L.M. Silverman (1981). "Topology induced by the Hankel norm in the space of transfer matrices," *Proc. CDC.*

Jonckheere, E.A. and J.C. Juang (1986). "Toeplitz + Hankel structures in H_∞ design and fast computation of achievable performance," Tech. Report, Dept. Elect. Eng., Univ. S. Cal., Los Angeles.

Khargonekar, P. and E. Sontag (1982). "On the relation between stable matrix fraction factorizations and regulable realizations of linear systems over rings," *IEEE Trans. Auto. Control*, vol. AC-27, pp. 627-638.

Khargonekar, P., K. Poolla, and A. Tannenbaum (1985). "Robust control of linear time-invariant plants using periodic compensation," *IEEE Trans. Auto. Cont.*, vol. AC-30, pp. 1088-1098.

Khargonekar, P. and A. Tannenbaum (1985). "Noneuclidean metrics and the robust stabilization of systems with parameter uncertainty," *IEEE Trans. Auto. Cont.*, vol. AC-30, pp. 1005-1013.

Khargonekar, P. and K. Poolla (1986a). "Robust stabilization of distributed systems," *Automatica*, vol. 22, pp. 77-84.

Khargonekar, P. and K. Poolla (1986b). "Uniformly optimal control of linear time-invariant plants: nonlinear time-varying controllers," *Systems and Control Letters*, vol. 6, pp. 303-308.

Kimura, H. (1984). "Robust stabilization for a class of transfer functions," *IEEE Trans. Auto. Cont.*, vol. AC-29, pp. 788-793.

Kimura, H. (1986). "On interpolation-minimization problem in H_∞," *Control--Theory and Advanced Technology*, vol. 2, pp. 1-25.

Kung, S.Y. and D.W. Lin (1981). "Optimal Hankel-norm model reductions: multivariable systems," *IEEE Trans. Auto. Cont.*, vol. AC-26, pp. 832-852.

Kwakernaak, H. (1984). "Minimax frequency-domain optimization of multivariable linear feedback systems," *IFAC World Congress*, Budapest.

Kwakernaak, H. (1985). "Minimax frequency domain performance and robustness optimization of linear feedback systems," *IEEE Trans. Auto. Cont.*, vol. AC-30, pp. 994-1004.

Kwakernaak, H. (1986). "A polynomial approach to minimax frequency domain optimization of multivariable systems," *Int. J. Control*, vol. 44, pp. 117-156.

Limebeer, D.J.N. and Y.S. Hung (1986). "An analysis of the pole-zero cancellations in H_∞ optimal control problems of the first kind," Technical Report , Dept. Elect. Eng., Imperial College, London.

Minto, K.D. (1985). "Design of reliable control systems: theory and computations," Ph.D. Thesis, Dept. Elect. Eng., Univ. Waterloo, Waterloo, Canada.

Nehari, Z. (1957). "On bounded bilinear forms," *Ann. of Math.*, vol. 65, pp. 153-162.

Nett, C.N., C.A. Jacobson, and M.J. Balas (1984). "A connection between state-space and doubly coprime fractional representations," *IEEE Trans. Auto. Cont.*, vol. AC-29, pp. 831-832.

Nett, C.N. (1985). "Algebraic aspects of linear control system stability," *Proc. CDC.*

Nikolskii, N.K. (1986). *Treatise on the Shift Operator,* Springer-Verlag, Berlin.

O'Young, S. (1985). "Performance trade-offs in the design of multivariable controllers," Ph. D. Thesis, Dept. Elect. Eng., Univ. Waterloo, Waterloo, Canada.

O'Young, S. and B.A. Francis (1985). "Sensitivity trade-offs for multivariable plants," *IEEE Trans. Auto. Cont.*, vol. AC-30, pp. 625-632.

O'Young, S. and B.A. Francis (1986). "Optimal performance and robust stabilization," *Automatica*, vol. 22, pp. 171-183.

Paley, R.E.A.C. and N. Wiener (1934). *Fourier Transforms in the Complex Domain,* AMS, Providence, R.I..

Pandolfi, L. and A.W. Olbrot (1986). "On the minimization of sensitivity to additive disturbances for linear-distributed parameter MIMO feedback systems," *Int. J. Control*, vol. 43, pp. 389-399.

Pandolfi, L. (1986). "Some inequalities for nonminimum phase systems," Tech. Report no. 3, Dip. di Matematica, Politecnico di Torino, Torino, Italy.

Pascoal, A. and P. Khargonekar (1986). "Remarks on weighted sensitivity minimization," Tech. Rept., Dept. EE, Univ. Minn., Minneapolis.

Pernebo, L. (1981). "An algebraic theory for the design of controllers for linear multivariable systems, parts I & II," *IEEE Trans. Auto. Cont.*, vol. AC-26, pp. 171-194.

Postlethwaite, I. and Y.K. Foo (1985). "All solutions, all-pass form solutions, and the 'best' solutions to an H_∞ optimization problem in robust control," *Proc. Sym. MTNS*, Stockholm.

Power, S.C. (1982). *Hankel Operators on Hilbert Space,* Pitman, London.

Rosenblum, M. and J. Rovnyak (1985). *Hardy Classes and Operator Theory,* Oxford Univ. Press, New York.

Rudin, W. (1966). *Real and Complex Analysis,* McGraw-Hill, New York.

Safonov, M. and M. Athans (1981). "A multiloop generalization of the circle criterion for stability margin analysis," *IEEE Trans. Auto. Cont.*, vol. AC-26, pp. 415-422.

Safonov, M.G. and B.S. Chen (1982). "Multivariable stability margin optimization with decoupling and output regulation," *IEE Proc., Part D*, vol. 129, pp. 276-282.

Safonov, M.G. (1983). "L_∞-optimal sensitivity versus stability margin," *Proc. CDC.*

Safonov, M.G. (1985). "Optimal diagonal scaling for infinity norm optimization," *Proc. ACC*, Boston.

Safonov, M.G. and M.S. Verma (1985). "L_∞ sensitivity optimization and Hankel approximation," *IEEE Trans. Auto. Cont.*, vol. AC-30, pp. 279-280.

Sarason, D. (1967). "Generalized interpolation in H_∞," *Trans. AMS*, vol. 127, pp. 179-203.

Sideris, A. and M.G. Safonov (1985). "Design of linear control systems for robust stability and performance," *Proc. IFAC Workshop on Model Error Concepts and Compensation*, Boston.

Silverman, L. and M. Bettayeb (1980). "Optimal approximation of linear systems," *Proc. JACC.*

Sz.-Nagy, B. and C. Foias (1970). *Harmonic Analysis of Operators on Hilbert Space,* North-Holland.

Tannenbaum, A. (1977). "On the blending problem and parameter uncertainty in control theory," Tech. Report, Dept. Math., Weizmann Inst. Sci., Rehovot, Israel.

Tannenbaum, A. (1980). "Feedback stabilization of linear dynamical plants with uncertainty in the gain factor," *Int. J. Control,* vol. 32, pp. 1-16.

Tannenbaum, A. (1982). "Modified Nevanlinna-Pick interpolation of linear plants with uncertainty in the gain factor," *Int. J. Control,* vol. 36, pp. 331-336.

Verma, M. and E. Jonckheere (1984). "L_∞-compensation with mixed sensitivity as a broadband matching problem," *Systems and Control Letters,* vol. 4, pp. 125-130.

Verma, M.S. (1985). "Synthesis of infinity-norm optimal linear feedback systems," Ph. D. Thesis, Dept. Elect. Eng., USC, Los Angeles, CA.

Verma, M.S. (1986). "Robust stabilization of linear time-invariant systems," Technical Report, Dept. Elect. Comp. Eng., Univ. of Texas, Austin, TX.

Vidyasagar, M. (1972). "Input-output stability of a broad class of linear time-invariant multivariable feedback systems," *SIAM J. Control,* vol. 10, pp. 203-209.

Vidyasagar, M. (1984). "The graph metric for unstable plants and robustness estimates for feedback stability," *IEEE Trans. Auto. Cont.,* vol. AC-29, pp. 403-418.

Vidyasagar, M. (1985a). *Control System Synthesis: A Factorization Approach,* MIT Press, Cambridge, MA.

Vidyasagar, M. (1985b). "Filtering and robust regulation using a two-parameter controller," *Proc. Sym. MTNS*, Stockholm.

Vidyasagar, M. and H. Kimura (1986). "Robust controllers for uncertain linear multivariable systems," *Automatica*, vol. 22, pp. 85-94.

Wang, Z.Z. and J.B. Pearson (1984). "Regulation and optimal error reduction in linear multivariable systems," *Proc. IFAC World Congress*, Budapest.

Wonham, W.M. (1985). *Linear Multivariable Control*, Springer-Verlag, New York.

Youla, D.C. (1961). "On the factorization of rational matrices," *IRE Trans. Inf. Theory*, vol. IT-7, pp. 172-189.

Youla, D.C., H.A. Jabr, and J.J. Bongiorno Jr. (1976). "Modern Wiener-Hopf design of optimal controllers: part II," *IEEE Trans. Auto. Cont.*, vol. AC-21, pp. 319-338.

Young, N.J. (1986a). "The Nevanlinna-Pick problem for matrix-valued functions," *J. Operator Theory*. To appear.

Young, N.J. (1986b). "An algorithm for the super-optimal sensitivity-minimising controller," *Proc. Workshop on New Perspectives in Industrial Control System Design using* H_∞ *Methods*, Oxford.

Zames, G. (1960). "Nonlinear operators for system analysis," Tech. Report 370, Research Lab. of Electronics, MIT, Cambridge, MA.

Zames, G. (1976). "Feedback and complexity," Special plenary lecture addendum, IEEE Conf. Dec. Control.

Zames, G. (1979). "Optimal sensitivity and feedback: weighted seminorms, approximate inverses, and plant invariant schemes," *Proc. Allerton Conf.*.

Zames, G. (1981). "Feedback and optimal sensitivity: model reference transformations, multiplicative seminorms, and approximate inverses," *IEEE Trans. Auto. Cont.*, vol. AC-23, pp. 301-320.

Zames, G. and B.A. Francis (1983). "Feedback, minimax sensitivity, and optimal robustness," *IEEE Trans. Auto. Cont.*, vol. AC-28, pp. 585-601.

Zames, G. and D. Bensoussan (1983). "Multivariable feedback, sensitivity, and decentralized control," *IEEE Trans. Auto. Cont.*, vol. AC-28, pp. 1030-1035.

Lecture Notes in Control and Information Sciences

Edited by M. Thoma

Vol. 43: Stochastic Differential Systems
Proceedings of the 2nd Bad Honnef Conference
of the SFB 72 of the DFG at the University of Bonn
June 28 – July 2, 1982
Edited by M. Kohlmann and N. Christopeit
XII, 377 pages. 1982.

Vol. 44: Analysis and Optimization of Systems
Proceedings of the Fifth International
Conference on Analysis and Optimization of Systems
Versailles. December 14–17, 1982
Edited by A. Bensoussan and J. L. Lions
XV, 987 pages, 1982

Vol. 45: M. Arató
Linear Stochastic Systems
with Constant Coefficients
A Statistical Approach
IX, 309 pages. 1982

Vol. 46: Time-Scale Modeling of Dynamic Networks
with Applications to Power Systems
Edited by J. H. Chow
X, 218 pages. 1982

Vol. 47: P. A. Ioannou, P. V. Kokotovic
Adaptive Systems with Reduced Models
V, 162 pages. 1983

Vol. 48: Yaakov Yavin
Feedback Strategies for Partially
Observable Stochastic Systems
VI, 233 pages, 1983

Vol. 49: Theory and Application of Random Fields
Proceedings of the IFIP-WG 7/1
Working Conference
held under the joint auspices of the
Indian Statistical Institute
Bangalore, India, January 1982
Edited by G. Kallianpur
VI. 290 pages. 1983

Vol. 50: M. Papageorgiou
Applications of Automatic Control Concepts
to Traffic Flow Modeling and Control
IX, 186 pages. 1983

Vol. 51: Z. Nahorski, H.F. Ravn, R.V.V. Vidal
Optimization of Discrete Time Systems
The Upper Boundary Approach
V, 137 pages 1983

Vol. 52: A. L. Dontchev
Perturbations, Approximations and Sensitivity Analysis
of Optimal Control Systems
IV, 158 pages. 1983

Vol. 53: Liu Chen Hui
General Decoupling Theory of Multivariable
Process Control Systems
XI, 474 pages. 1983

Vol. 54: Control Theory for Distributed
Parameter Systems and Applications
Edited by F. Kappel, K. Kunisch,
W. Schappacher
VII, 245 pages. 1983.

Vol. 55: Ganti Prasada Rao
Piecewise Constant Orthogonal Functions
and Their Application to Systems and Control
VII, 254 pages. 1983.

Vol. 56: Dines Chandra Saha, Ganti Prasada Rao
Identification of Continuous
Dynamical Systems
The Poisson Moment Functional
(PMF) Approach
IX, 158 pages. 1983.

Vol. 57: T. Söderström, P. G. Stoica
Instrumental Variable Methods
for System Identification
VII, 243 pages. 1983.

Vol. 58: Mathematical Theory of
Networks and Systems
Proceedings of the MTNS-83 International
Symposium
Beer Sheva, Israel, June 20–24, 1983
Edited by P. A. Fuhrmann
X, 906 pages. 1984

Vol. 59: System Modelling and Optimization
Proceedings of the 11th IFIP Conference
Copenhagen, Denmark, July 25-29, 1983
Edited by P. Thoft-Christensen
IX, 892 pages. 1984

Vol. 60: Modelling and Performance
Evaluation Methodology
Proceedings of the International Seminar
Paris, France, January 24–26, 1983
Edited by F. Bacelli and G. Fayolle
VII, 655 pages. 1984

Vol. 61: Filtering and Control of Random
Processes
Proceedings of the E.N.S.T.-C.N.E.T. Colloquium
Paris, France, February 23–24, 1983
Edited by H. Korezlioglu, G. Mazziotto, and
J. Szpirglas
V, 325 pages. 1984

Lecture Notes in Control and Information Sciences

Edited by M. Thoma and A. Wyner

Vol. 62: Analysis and Optimization of Systems
Proceedings of the Sixth International Conference on Analysis and Optimization of Systems
Nice, June 19–22, 1984
Edited by A. Bensoussan, J. L. Lions
XIX, 591 pages. 1984.

Vol. 63: Analysis and Optimization of Systems
Proceedings of the Sixth International Conference on Analysis and Optimization of Systems
Nice, June 19–22, 1984
Edited by A. Bensoussan, J. L. Lions
XIX, 700 pages. 1984.

Vol. 64: Arunabha Bagchi
Stackelberg Differential Games in Economic Models
VIII, 203 pages, 1984

Vol. 65: Yaakov Yavin
Numerical Studies in Nonlinear Filtering
VIII, 273 pages, 1985.

Vol. 66: Systems and Optimization
Proceedings of the Twente Workshop
Enschede, The Netherlands, April 16–18, 1984
Edited by A. Bagchi, H. Th. Jongen
X, 206 pages, 1985.

Vol. 67: Real Time Control of Large Scale Systems
Proceedings of the First European Workshop
University of Patras, Greece, Juli 9–12, 1984
Edited by G. Schmidt, M. Singh, A. Titli, S. Tzafestas
XI, 650 pages, 1985.

Vol. 68: T. Kaczorek
Two-Dimensional Linear Systems
IX, 397 pages, 1985.

Vol. 69: Stochastic Differential Systems – Filtering and Control
Proceedings of the IFIP-WG 7/1 Working Conference
Marseille-Luminy, France, March 12-17, 1984
Edited by M. Metivier, E. Pardoux
X, 310 pages, 1985.

Vol. 70: Uncertainty and Control
Proceedings of a DFVLR International Colloquium
Bonn, Germany, March, 1985
Edited by J. Ackermann
IV, 236 pages, 1985.

Vol. 71: N. Baba
New Topics in Learning Automata Theory and Applications
VII, 231 pages, 1985.

Vol. 72: A. Isidori
Nonlinear Control Systems: An Introduction
VI, 297 pages, 1985.

Vol. 73: J. Zarzycki
Nonlinear Prediction Ladder-Filters for Higher-Order Stochastic Sequences
V, 132 pages, 1985.

Vol. 74: K. Ichikawa
Control System Design based on Exact Model Matching Techniques
VII, 129 pages, 1985.

Vol. 75: Distributed Parameter Systems
Proceedings of the 2nd International Conference, Vorau, Austria 1984
Edited by F. Kappel, K. Kunisch, W. Schappacher
VIII, 460 pages, 1985.

Vol. 76: Stochastic Programming
Edited by F. Archetti, G. Di Pillo, M. Lucertini
V, 285 pages, 1986.

Vol. 77: Detection of Abrupt Changes in Signals and Dynamical Systems
Edited by M. Basseville, A. Benveniste
X, 373 pages, 1986.

Vol. 78: Stochastic Differential Systems
Proceedings of the 3rd Bad Honnef Conference, June 3–7, 1985
Edited by N. Christopeit, K. Helmes, M. Kohlmann
V, 372 pages, 1986.

Vol. 79: Signal Processing for Control
Edited by K. Godfrey, P. Jones
XVIII, 413 pages, 1986.

Vol. 80: Artificial Intelligence and Man-Machine Systems
Edited by H. Winter
IV, 211 pages, 1986.